IEEE Guide for the Installation of Electrical Equipment to Minimize Electrical Noise Inputs to Controllers from External Sources

Published by
The Institute of Electrical and Electronics Engineers, Inc

Distributed in cooperation with
Wiley-Interscience, a division of John Wiley & Sons, Inc

IEEE
Std 518-1982
(Revision of
ANSI/IEEE Std 518-1977)

IEEE Guide for the Installation of Electrical Equipment to Minimize Electrical Noise Inputs to Controllers from External Sources

Sponsor

Industrial Control Committee of the
IEEE Industrial Applications Society

ISBN 0-471-89359-5

Library of Congress Catalog Number 82-083305

October 1, 1982

SH08813

IEEE Standards documents are developed within the Technical Committees of the IEEE Societies and the Standards Coordinating Committees of the IEEE Standards Board. Members of the committees serve voluntarily and without compensation. They are not necessarily members of the Institute. The standards developed within IEEE represent a consensus of the broad expertise on the subject within the Institute as well as those activities outside of IEEE which have expressed an interest in participating in the development of the standard.

Use of an IEEE Standard is wholly voluntary. The existence of an IEEE Standard does not imply that there are no other ways to produce, test, measure, purchase, market, or provide other goods and services related to the scope of the IEEE Standard. Furthermore, the viewpoint expressed at the time a standard is approved and issued is subject to change brought about through developments in the state of the art and comments received from users of the standard. Every IEEE Standard is subjected to review at least once every five years for revision or reaffirmation. When a document is more than five years old, and has not been reaffirmed, it is reasonable to conclude that its contents, although still of some value, do not wholly reflect the present state of the art. Users are cautioned to check to determine that they have the latest edition of any IEEE Standard.

Comments for revision of IEEE Standards are welcome from any interested party, regardless of membership affiliation with IEEE. Suggestions for changes in documents should be in the form of a proposed change of text, together with appropriate supporting comments.

Interpretations: Occasionally questions may arise regarding the meaning of portions of standards as they relate to specific applications. When the need for interpretations is brought to the attention of IEEE, the Institute will initiate action to prepare appropriate responses. Since IEEE Standards represent a consensus of all concerned interests, it is important to ensure that any interpretation has also received the concurrence of a balance of interests. For this reason IEEE and the members of its technical committees are not able to provide an instant response to interpretation requests except in those cases where the matter has previously received formal consideration.

Comments on standards and requests for interpretations should be addressed to:

Secretary, IEEE Standards Board
345 East 47th Street
New York, NY 10017
USA

Foreword

(This Foreword is not a part of IEEE Std 518-1982, IEEE Guide for the Installation of Electrical Equipment to Minimize Electrical Noise Inputs to Controllers from External Sources.)

The rapidly expanding use of computers and solid-state controllers in industry requires consideration of many factors not previously important in the design, installation, and operation of other forms of controllers. Many portions of the control circuits of computers and solid-state controllers designed for operation at low-energy low-voltage signal levels are susceptible to disturbances by excessive electrical noise. Erratic controller operation may result unless suitable precautions are taken. The following recommendations are intended as an installation guide for industrial controls involving low-energy-level equipment to minimize electrical noise inputs from external sources.

The electrical noise guide is comprised of six sections. Sections 1 and 2 state the scope and service conditions. Sections 3 through 5 provide the technical foundation for the recommendations given in Section 6. Section 6 is intended to stand alone as the working section of the electrical noise guide.

At the time it approved this guide, the Industrial Control Systems Subcommittee of the Industrial Control Committee had the following membership:

George W. Younkin, *Chairman*

Pritindra Chowdhuri
Jim Feltner
Carl W. Kellenbenz
Dale H. Levisay
Edward J. Louma
John Riley
Heinz M. Schlicke
R.M. Showers
Merle R. Swinehart
James R. Wilson

The Industrial Control Systems Subcommittee had the following liaison members:

R.S. Burns
L.M. Johnson
George E. Heidenreich
J.L. Koepfinger
P.C. Lyons
F.D. Martzloff
R.M. Morris
C.H. Moser
F. Oettinger
R.D. Smith
Thomas R. Thompson
F.W. Wells

Acknowledgment

Figures 45, 46, 47, 48, 49, 50, 55, 61, 62 and 63 are reprinted, with permission, from Dr. Heinz Schlicke's book, *Electromagnetic Compossibility (Applied Principles of Cost-Effective Control of Electro-magnetic Interference and Hazards)*, second enlarged edition, Marcel Dekker, New York, 1982.

Contents

FIGURES PAGE

FIGURES PAGE

FIGURES PAGE

TABLES

IEEE Guide for the Installation of Electrical Equipment to Minimize Electrical Noise Inputs to Controllers from External Sources

1. Scope

The purpose of this document is to develop a guide for the installation and operation of industrial controllers to minimize the disturbing effects of electrical noise on these controllers.

2. Service Conditions

This guide is limited to techniques for the installation of controllers and control systems so that proper operation can be achieved in the presence of electrical noise. This guide is not intended to be a guide for the internal design of electrical controllers or for the prevention of the generation of electrical noise resulting from their operation.

3. Identification of Electrical Noise in Control Circuits

3.1 Definition of Noise (for the purpose of this guide).
electrical noise. Unwanted electrical signals, which produce undesirable effects in the circuits of the control systems in which they occur.

3.2 Significance of Electrical Noise. In any process under control there are at least two variables.

(1) The *controlled* variable is a condition or characteristic that can be measured, correlated to the process, and subsequently controlled. The con-

trolled variable may be temperature, pressure, liquid level, speed, or another similar physical quantity.

(2) The *manipulated* variable is the electrical power, energy, or material supplied to the process to keep the process under control.

The control system includes all the devices and equipment to monitor the controlled variable and to adjust the manipulated variable to keep the process at the desired condition.

The signal flow in a control system is shown in Fig 1. A primary element of some sort detects the condition or magnitude of the controlled variable and produces an electrical signal. These measurement signals (1) are directed to the *controller*, which accepts the signal, modifies or amplifies it, or does both, in relation to the constraints (2) which define the range of the controlled variable directly or indirectly. The controller output signal (3) is directed to the final control elements, which receive the signal and, in response to it, adjust the manipulated variable to maintain the process at the desired value or within the allowable tolerance. The complexity of the controller and the associated control system is dependent on the process characteristics.

In early control systems the controller function was performed by a person who was capable of using judgment in maintaining the process under control. As the state of the art progressed, it was possible to design more functional control systems so that humans could be removed from the control loop. The required precision in industrial processes, coupled with advances in electronic devices, has led to the development of the all-electronic solid-state controller. These controllers, which are usually more complex, may operate at very low power levels. Electrical disturbances are more likely to affect them because of the lower power levels. In addition to this, the system must operate at high speeds in order to provide the required speed of

Fig 1
Signal Flow in Control System

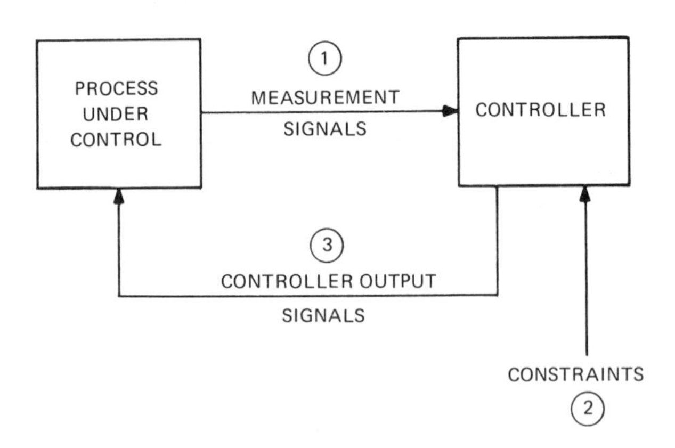

response. These factors tend to make noise a more important factor in the operation of the system.

If the definition and concept of electrical noise in control circuits is examined, it is possible to divide the problem into three subdivisions:

(1) Sources of electrical noise (see 3.3)
(2) Coupling of noise from the source to the control circuit (see 3.4)
(3) Susceptibility of the control circuit to electrical noise (see 3.5)

3.3 Sources of Electrical Noise. Typical sources of electrical noise are electrical or electromechanical devices which cause or produce very rapid or large amplitude changes in voltage or current. A partial list of the devices known to be sources of noise is given below:

(1) Switches operating inductive loads
(2) Thyristors or other semiconductors used in the switching mode
(3) Welding machines
(4) Heavy current conductors
(5) Fluorescent lights
(6) Thyratron and ignitron tubes
(7) Neon lights

There are other sources of noise which appear in a control circuit due to the method of connecting the control wires. In general these noise sources are in series with the control signal and can be controlled by careful attention to connections. Some of these sources are:

(1) Thermal voltages between dissimilar metals
(2) Chemical voltages due to electrolyte between poorly connected leads
(3) Thermal noise of resistors

3.4 Coupling of Electrical Noise. If some type of electrical coupling exists between the source of the noise voltage or current and one or more wires or elements of the control circuit, a voltage will appear in the control circuit. The four possible types of coupling are discussed in this section. (Whether the voltage that appears in the control circuit actually results in noise, as defined in this guide, is dependent upon the susceptibility of the control circuit; see 3.5.)

3.4.1 Common Impedance Coupling. This type of noise can produce problems any time two circuits share any common conductors or impedances (even in the case of common power sources in extreme cases). Probably one of the most usual methods of common impedance coupling is the use of long common neutral or ground wires. Figure 2 shows an example of this in a circuit which is described as *single ended* because the signal voltage appears as a voltage between a single wire and the neutral. In Fig 2 two amplifiers AR1 and AR2 are used to amplify the output of an iron-constantan thermocouple TC. A load operated by contact 1a is also associated with the circuit. The lead A-B is 50 ft of AWG No 12 wire, which has a resistance of 1.6 Ω per 1000 ft. Thus

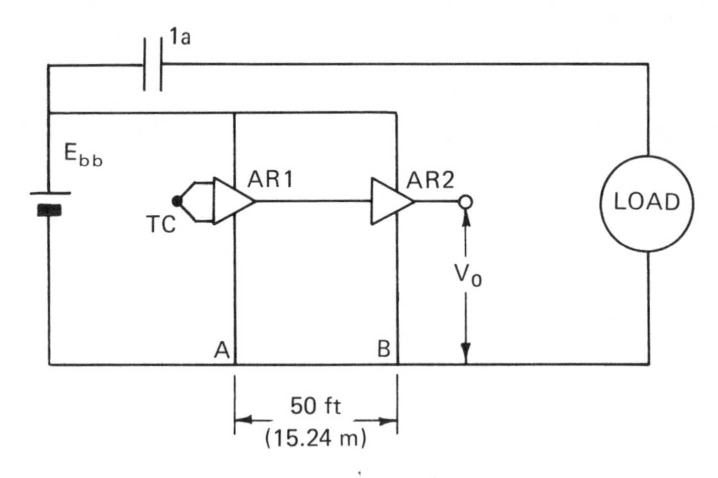

Fig 2
Single-Ended Circuit with Long Common Wire

$$R_{AB} = \frac{50}{1000} \cdot 1.6 = 0.08\ \Omega$$

If the load requires 0.5 A from the source E_{bb}, the voltage between points A and B will be

$$V_{AB} = 0.5 \cdot 0.08 = 0.04\ \text{V}$$

Assume that the amplifier AR1 close to the thermocouple TC will amplify by 100. This will mean that 0.04 V from A to B will appear to AR2 just like a 0.04/100 V (or 0.4 mV) change in the input voltage to AR1. For an iron-constantan thermocouple this would be equivalent to a 14 °F temperature change near 100 °F.

When the current through the load is interrupted by contact 1a, it could change from 0.5 A to 0 in less than 1 μs. With this rapid change in current in line A-B the effect of the self-inductance must also be considered. Normally inductance is a property attached to a complete circuit, but it is possible to attach a value of inductance to teach segment making up a circuit. This results in a way of calculating the voltage induced in a portion of a circuit (such as line A-B) by a change in current in the circuit. The formula for the self-inductance of a straight wire at high frequency is derived in [1] [1]:

$$L = 0.002l \left[\log_e \frac{2l}{r} - \frac{3}{4}\right]\ \mu\text{H}$$

[1] The numbers in brackets correspond to those of the references listed in 3.6.

where l is the length of wire in centimeters and r is the radius of the conductor in centimeters. For wire A–B in Fig 2 this would be

$$l = 1524 \text{ cm}$$
$$r = 0.205 \text{ cm}$$

$$L = 3.05 \left[\log_e \frac{2 \cdot 1524}{0.205} - 0.75 \right] \quad \mu\text{H}$$

$$= 3.05 \left[9.6 - 0.75 \right] \quad \mu\text{H}$$

$$= 3.05 \cdot 8.85 = 27 \quad \mu\text{H}$$

When the current through the load is interrupted, it could change from 0.5 A to 0 in 1 μs, and the voltage between points A and B becomes

$$V_{AB} = L \frac{\Delta i}{\Delta t}$$

$$= 27 \cdot 10^{-6} \cdot \frac{0.5}{10^{-6}} = 13.5 \text{ V peak}$$

This disturbance is over in a short time, but it could cause trouble in a control circuit.

Many times it is possible to recognize obvious common impedance paths in a control circuit and eliminate them. A good example is to use two distinct wires for each transducer feeding into a control element. However, there are other common impedance paths which are not as obvious. Some of these can be listed to present a better understanding of this problem:

(1) Common power supply
(2) Inductance of capacitors at high frequency
(3) Inductance of lead wires
(4) Stray capacitance from the enclosure to ground
(5) Inductance of busbars on the common or neutral in a building
(6) Resistance of the earth or ground

3.4.2 Magnetic Coupling. This type of coupling is often called inductive because it is proportional to the mutual inductance between a control circuit and a source of interference current. Its magnitude also depends upon the rate of change of the interference current. This type of coupling does not depend upon any conductive coupling between the noise source and the control circuit. Figure 3 illustrates the nature of magnetic coupling.

In the circuit of Fig 3 assume that all four wires, A, B, C, and D are placed in the same nonmetallic raceway with a separation of 3 in between A and B, 6 in between B and C, and 1 in between C and D. Under these conditions the mutual inductance between the noise-producing circuit (A and B) and the

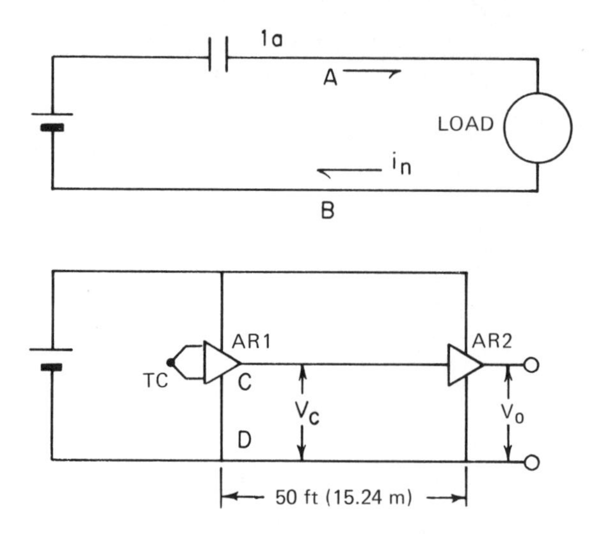

Fig 3
Circuit Showing Magnetic Coupling

control circuit (C and D) can be calculated from [2], p 57, eq (2-107). For the given distances

A		B		C		D
o	←3 in→ (7.62 cm)	o	←6 in→ (15.24 cm)	o	←1 in→ (2.54 cm)	o

we have

$$M_{nc} = \frac{140.4}{10^9} \cdot l \cdot \log_{10}\left[\frac{d_{AC} \cdot d_{BD}}{d_{BC} \cdot d_{AD}}\right] \quad \mu H$$

where l is the length of wire in feet and d is the distance between wires (any units may be used). Thus

$$M_{nc} = 7020 \cdot 10^{-9} \cdot \log_{10}\left[\frac{9}{6} \cdot \frac{7}{10}\right] \quad \mu H$$

$$= 0.149 \ \mu H$$

If the rate of current decay $\Delta i_n / \Delta t$ is 0.5 A/μs, as assumed for the example in 3.4.1, the voltage induced in the control circuit would be

$$V_{CD} = M_{nc} \frac{\Delta i_n}{\Delta t}$$

$$= 0.149 \cdot 10^{-6} \cdot 0.5/10^{-6}$$

$$= 0.0745 \text{ V}$$

For an amplifier gain AR1 = 100, as assumed in the example in 3.4.1, this would be equivalent to a voltage at the thermocouple of 0.745 mV (26 °F at TC).

3.4.3 Electrostatic Coupling. This type of coupling is often called capacitive coupling because it is proportional to the capacitance between a control lead and a source of interference or noise voltage. Its magnitude depends upon the rate of change of the noise voltage and the impedance between elements of the noise circuit and elements of the control circuit.

The same circuit as that used to illustrate magnetic coupling can be used to illustrate electrostatic coupling (Fig 4). However, the voltage between the wires and the capacitance is now important. The formula for the capacitance between two parallel wires in air is given by the following formula from [2], p 65, eq (2-144), modified as noted in sec 2-130:

$$C = \frac{3.68\ \epsilon_R}{\log_{10}\left(\frac{2s}{d} - \frac{d}{2s}\right)} \text{ pF/ft}$$

where d is the diameter of the wires (0.08 in for AWG No 12), s is the distance between wires (in the same units as d), and ϵ_R is the relative permittivity of the medium between conductors. (For this example it is sufficient to assume the relative permittivity to be that of air, which is 1. In practice, the equivalent permittivity will be somewhere between 1 and 3, depending on the distribution and type of insulation on the wire.) For the given distances

A		B		C		D
o	←3 in→ (7.62 cm)	o	←6 in→ (15.24 cm)	o	←1 in→ (2.54 cm)	o

we have

$$C_{AD} = \frac{3.68 \cdot 50}{\log_{10}\left(\frac{20}{0.08} - \frac{0.08}{20}\right)} = \frac{184}{2.398} = 76.7 \text{ pF}$$

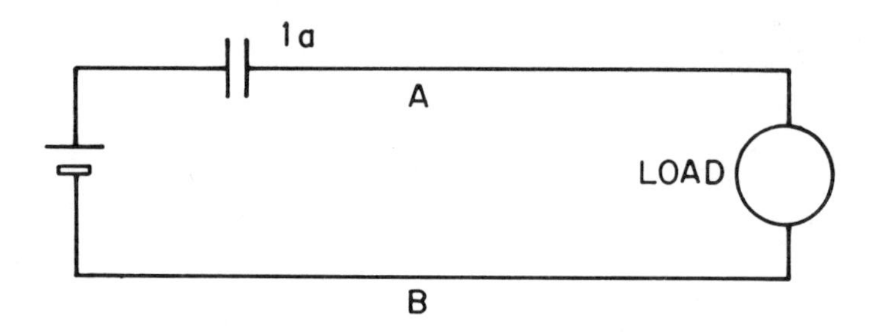

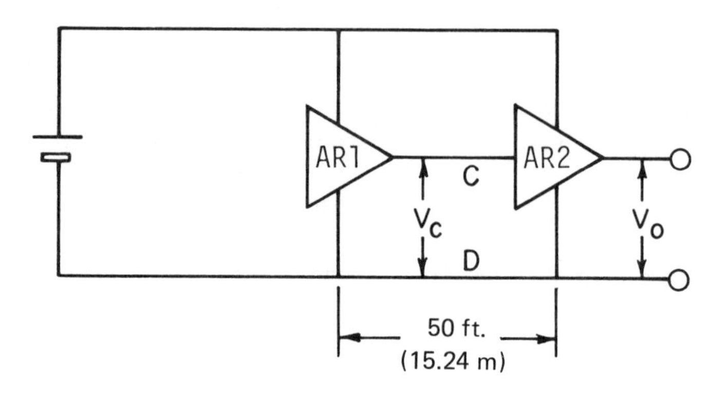

Fig 4
Circuit Showing Electrostatic Coupling

$$C_{AC} = \frac{3.68 \cdot 50}{\log_{10}\left(\frac{18}{0.08} - \frac{0.08}{18}\right)} = \frac{184}{2.351} = 78.3 \text{ pF}$$

$$C_{BD} = \frac{3.68 \cdot 50}{\log_{10}\left(\frac{14}{0.08} - \frac{0.08}{14}\right)} = \frac{184}{2.242} = 82.1 \text{ pF}$$

$$C_{BC} = \frac{3.68 \cdot 50}{\log_{10}\left(\frac{12}{0.08} - \frac{0.08}{12}\right)} = \frac{184}{2.176} = 84.6 \text{ pF}$$

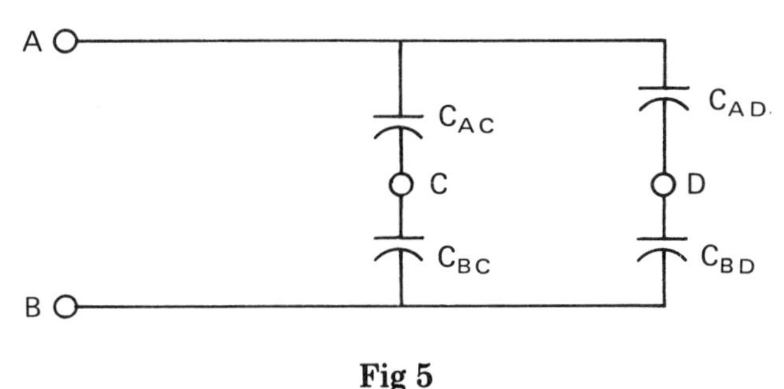

Fig 5
Equivalent Circuit

These are the important capacitors in this case, but it must be realized that all objects have capacitance to all other objects resulting at times in unexpected *sneak* circuits. For instance, an isolated oscilloscope used for troubleshooting might have a capacitance to a conducting grid in a concrete floor of 150 pF.

When interrupting the current to the load, it is possible for the voltage between lines A and B to rise to 2000 V. (This is representative of the showering arc voltage resulting from opening a relay coil with double break contacts.) The coupling mechanism between lines A–B and C–D is very complex, but the following simplistic analysis will illustrate a principle. Assume the equivalent circuit shown in Fig 5. Using the relationship of voltage division by capacitors,

$$V_1 = \frac{V_2 C_2}{C_1}$$

we have

$$V_{BC} = \frac{V_{AB}}{C_{BC}} \frac{C_{AC} \cdot C_{BC}}{C_{AC} + C_{BC}}$$

$$= \frac{2000}{84.6} \frac{78.3 \cdot 84.6}{78.3 + 84.6} = 961.32 \text{ V}$$

Similarly,

$$V_{BD} = \frac{V_{AB}}{C_{BD}} \frac{C_{AD} \cdot C_{BD}}{C_{AD} + C_{BD}}$$

$$= \frac{2000}{82.1} \frac{76.7 \cdot 82.1}{76.7 + 82.1} = 965.99 \text{ V}$$

and

$$V_{CD} = V_{BD} - V_{BC}$$

$$= 965.99 - 961.32 = 4.67 \text{ V}$$

Thus a considerable voltage can be induced in leads D and C, resulting in a voltage difference of 4.67 V at the input to amplifier AR2. This example was chosen to illustrate a serious condition. In practice this voltage difference would not be as great due to impedance paths to common, as explained in the following paragraph.

The process of electrostatically inducing a voltage in a lead wire can be summarized by the following equation (Fig 6). For this illustration assume rms voltages.

Fig 6

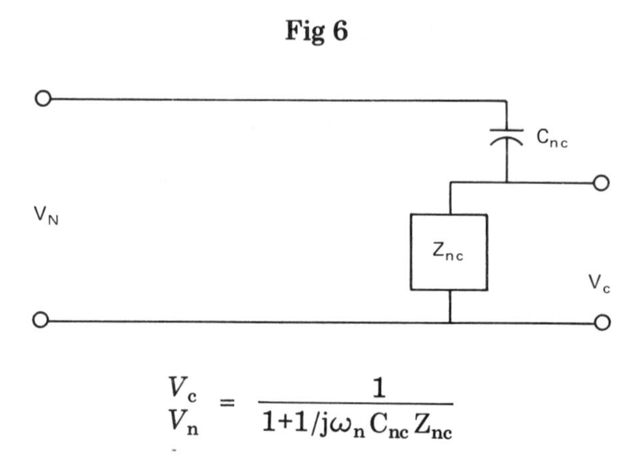

$$\frac{V_c}{V_n} = \frac{1}{1 + 1/j\omega_n C_{nc} Z_{nc}}$$

where

V_c = voltage induced in control wire
V_n = noise voltage source
ω_n = frequency of noise voltage
C_{nc} = capacitance between noise circuit and control wire
Z_{nc} = impedance between noise voltage reference and control signal wire

In order to keep the ratio V_c/V_n as low as possible, the product $\omega_n C_{nc} Z_{nc}$ must be kept as low as possible. Thus anything that can be done to reduce the coupling capacity C_{nc} or the impedance Z_{nc} will be helpful. A well-grounded shield on a control wire will do both.

3.4.4 Coupling by Electromagnetic Radiation. The voltages induced by magnetic or electrostatic induction are sometimes called near-field effects because this type of interference is produced close to interference sources. Further from the source, fields are associated with propagating radio waves or radiation fields. In general, radio waves are produced at distances greater than $\frac{1}{6}$ wavelength from the source of interference. For reference purposes the following chart can be used:

Frequency	1/6 Wavelength
1 MHz	1970 in (5000 cm)
10 MHz	197 in (500 cm)
100 MHz	19.7 in (50 cm)
1 GHz	1.97 in (5 cm)

This type of interference is particularly difficult to eliminate if present, because any shielding must be 100% complete, and any ground connection is only effective for immediately adjacent circuit elements. In most controllers this is not a problem. However, sources of these types of noise voltages, such as radio transmitters, high-frequency brazing equipment, and television stations, have been known to cause noise problems in control systems. The threshold of concern for ambient electromagnetic field strength will depend upon the level of susceptibility of the control system and whether it is susceptible directly to the incident continuous-wave signal, to a direct current which may be produced by rectification of the carrier, or to audio or video modulation frequencies which may appear as a result of demodulation in a system component. Whenever the field strength exceeds 1 V/m, a determination of susceptibility at the frequency of interest is advisable. This is the approximate field strength that would be produced by a 50 kW transmitter at a distance of 1.3 km (0.8 mi) or by a 5 W transmitter at 13.1 m (43 ft).

An approximation of the ambient field strength due to a known source can be calculated as

$$\mathrm{FS} = \frac{0.173\sqrt{P}}{D}$$

where

FS = field strength, in V/m
P = radiated power, in kW
D = distance from source, in km

This equation is based on free-space propagation, and in a specific case the actual field strength may be considerably more or less than that calculated by its use.

The coupling of electromagnetic radiation into a control circuit is via any incompletely shielded conductor, which will act as a receiving antenna for the signal. The received signal voltage will appear between the receiving conductor and ground. If the noise cannot be eliminated at the source by proper shielding, capacitors can be used to shunt these noise signals to ground, in many cases without reducing the performance of the system.

3.5 Susceptibility of Control Circuits. How well a control circuit can differentiate between noise and desired signals is a measure of its susceptibility. The susceptibility of a given control circuit is a function of the design of the circuit and can vary widely, even within a given class of devices. In general, high-power-level systems such as relay control systems have low susceptibility, while low-power-level systems such as those using integrated circuits have high susceptibility.

A control circuit is sensitive to a potential difference applied between two input terminals, whether that potential be noise voltage or a useful signal. Any voltage appearing between these input terminals is called a *normal-mode* voltage. Any voltage difference appearing between both terminals and some common point is called a *common-mode* voltage.

A major factor in the susceptibility of control and signal circuits is their ability to reject common-mode interference. Such interference may be introduced into a control signal circuit through inductive or capacitive coupling with the source of interference, or through electromagnetic radiation. Common-mode conversion is the process by which normal-mode interference is produced in a signal circuit by common-mode interference applied to the circuit. The amount of the common-mode conversion that occurs is a function of the balance of the signal circuit, that is, the electrical similarity (series impedance) of the two sides of the signal transmission path, and their symmetry with respect to a common reference plane, usually ground. For best balance, the requirement for symmetry must also be satisfied by the output of the signal source and the input of the signal receiver. These terminating impedances are generally determined by the equipment designer, not the user. However, the susceptibility of the system due to common-mode conversion may be reduced by judicious equipment selection and the use of balancing networks at equipment inputs and outputs.

Single-ended amplifiers have one of the input terminals at a common potential; they are thus more sensitive to noise. Differential amplifiers have both input terminals at some impedance level to ground. As such, any common-mode voltage (a voltage appearing on both input terminals simultaneously) will not normally result in an output signal (provided the impedance to noise voltage reference Z_{nc} is kept balanced for both input terminals). How well a differential amplifier rejects a common-mode voltage is usually indicated in amplifier specifications. Thus the use of differential amplifiers is a well-known method of decreasing circuit susceptibility when proper care is used in the design of input terminations.

Susceptibility is of concern to the user in the selection of the control system. However, once a control system has been chosen, the principles of noise reduction are the same, although the magnitude of the problem will depend upon the susceptibility of the chosen system. The installation of a control system must be made in such a way that noise sources present in the environment will not induce into the control circuit electrical signals of greater magnitude than the susceptibility of the circuit will permit.

3.6 References

[1] GROVER, F.W. *Inductance Calculations.* New York: Van Nostrand, 1944, p 35, eq (7).

[2] KNOWLTON, A.E., Ed. *Standard Handbook for Electrical Engineers,* 8th ed. New York: McGraw-Hill Book Co, 1949.

4. Classification

4.1 Electrical Noise Classification. Noise sources are listed here in the order of their importance. Noise will be characterized by *noise prints* (oscillograms of voltage and current versus time) or other data when available.

The voltage values listed in this section are intended to describe the characteristics of the particular phenomenon under discussion. The maximum voltage values listed could be limited to smaller values in any given system by the voltage breakdown of conductor spacing through air or over insulation surface [9].[2]

4.1.1 Showering Arc. The showering arc is caused by current chopping in electrical contacts. During circuit interruption, arc current will *chop* (suddenly cease to flow) at some value of current, which depends on the circuit constants, but is not less than the minimum arcing current [44]. This current chopping, in association with the circuit inductance (such as relay magnets) and shunt capacitance (such as coil self-capacitance and wiring capacitance), generates electrical noise transients, called showering arc [51], [70], where the peak voltages are a function of the contact gap breakdown, which in turn depends on the number and size of the contact gaps, contact material, atmosphere, and so on.

[2] The numbers in brackets correspond to those of the references listed in 4.6.

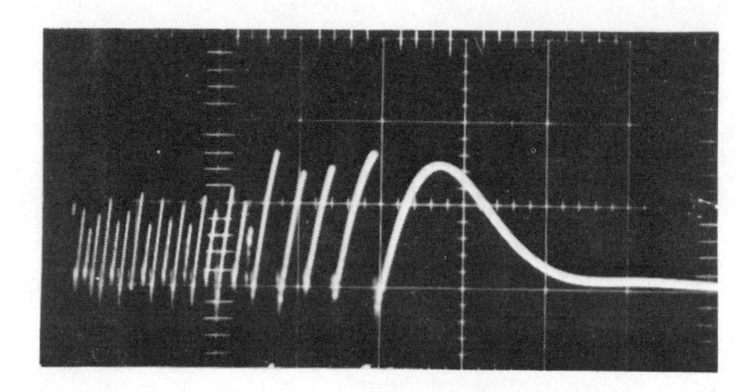

Fig 7
Showering Arc Voltage across Double Break Contacts Interrupting 120 V AC, 60 Hz Machine Tool Magnet

4.1.1.1 Electromagnetic Current Interruption. Typical noise prints, taken from [67], are shown in Figs 7-9 and in the following text. Each transient is unique. However, typical values of characteristics are as follows:

dv/dt (rise) up to 1000 V/μs
dv/dt (fall) $>$ 10 000 V/μs

(In [54] the fall is stated to be 2 000 000—6 000 000 V/μs.)

The repetition rate is up to 2 MHz. The duration is up to 1 ms during contact break, but will also occur during contact bounce on contact make.

Typical voltage peaks for a 120 V magnet are

1000 V for 1 contact gap in air
1500 V for 2 contact gaps in air
2500 V for 4 contact gaps in air
17000 V for 1 contact gap in vacuum

Typical voltage peaks for a 600 V magnet are 5600 V for 4 contact gaps in air

See 6.5.1.1 for methods for suppressing showering arc transients.

Most of the following noise sources are listed in outline form only. Information will be added as it becomes available in suitable form.

4.1.1.2 AC Motors

4.1.1.3 DC Motor Commutator

4.1.2 Solid-State Switching Devices. Power-line voltage transients can be initiated by switching power loads. For example, if a switch is closed to a resistive load when the voltage is not zero, the current will rise to the normal

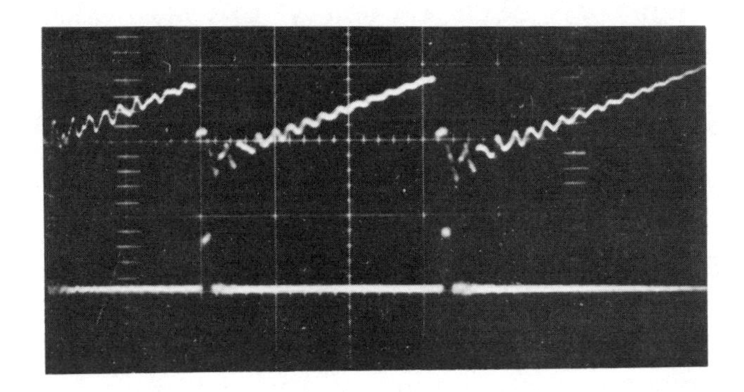

Fig 8
Upper Trace — Similar to Fig 7, but Expanded Time Scale;
Lower Trace — Current in Contacts
(Contacts connected by 100 ft, 300 Ω transmission line.)

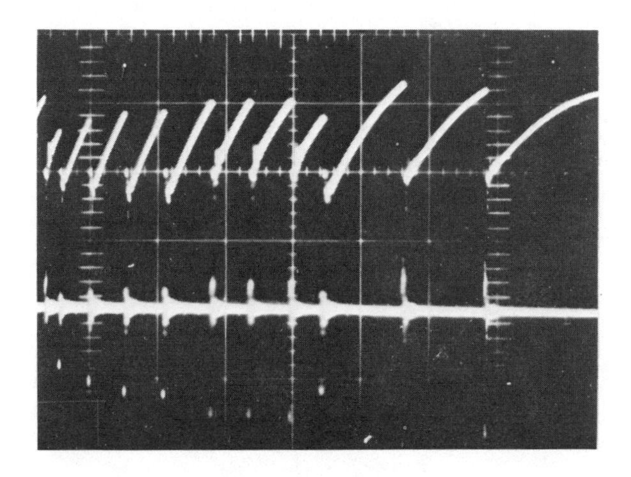

Fig 9
Upper Trace — Similar to Fig 7, but Expanded Time Scale;
Lower Trace — Voltage Induced in Adjacent 100 ft, 300 Ω Line
Terminated in Its Characteristic Impedance
(Note 300 V peaks coincident with Current Pulses in Fig 8.)

value as rapidly as the circuit parameters will permit. While this current is rising to its normal value, there will usually be a voltage transient associated with the expression

$$e = L\frac{\mathrm{d}i}{\mathrm{d}t} \quad V$$

where

L = circuit inductance.

While solid-state switching devices are limited to switching speeds less than those for hard contacts (manufacturers specify that the thyristor maximum permissible $\mathrm{d}i/\mathrm{d}t$ must initially be limited to approximately 10 A/μs), the total effect can be very disturbing because these devices are commonly used to control power by performing switching operations repeatedly, such as every half-cycle (commonly called phase control).

It is difficult to produce typical noise prints because of the wide variations in inductance, capacitance, and resistance, which affect the resultant transient due to switching; however, some examples follow.

4.1.2.1 Resistive Load. Phase control for lamp dimmers or heater control can be a source of noise. An example is shown in Figs 10–12.

4.1.2.2 Motor Load [65]. Semiconductor controlled rectifier (SCR) drives are capable of producing power-line noise. Figure 13 is a schematic diagram of an SCR control for a dc motor. In general, current flows through the dc motor from two of the three power lines selected by the SCRs to provide the proper motor terminal voltage. The SCR firing order is shown in the figure; this sequence is completed each cycle.

Fig 10
Noisy 30 A, 120 V, 60 Hz Laboratory Power Line
(The transients occurring just before voltage peak are caused by a 20 Ω resistive load phase controlled by a triac.)

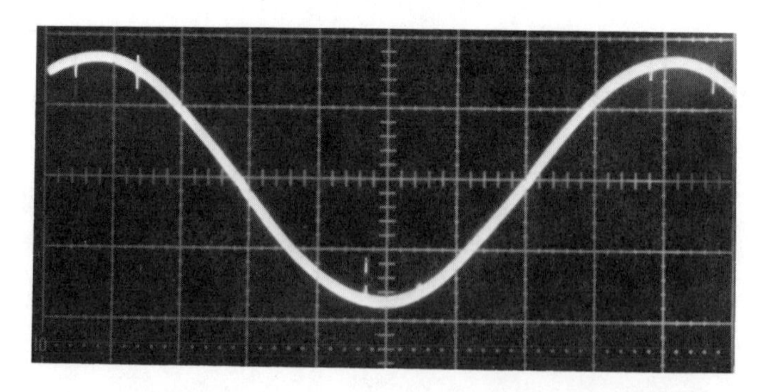

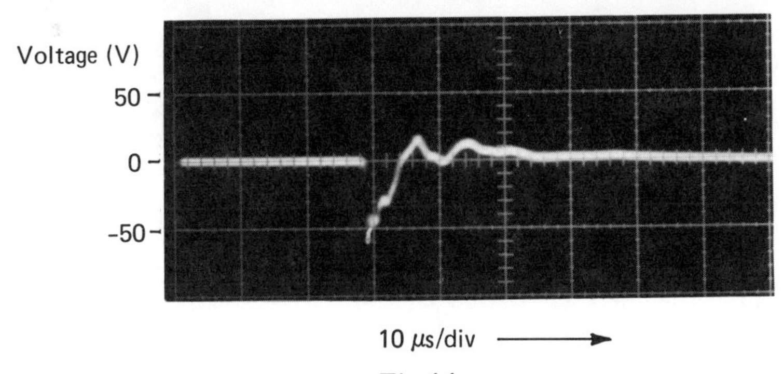

Fig 11
Same as Fig 10, but Expanded Time Scale and High-Pass Filter Used to Eliminate 60 Hz Voltage

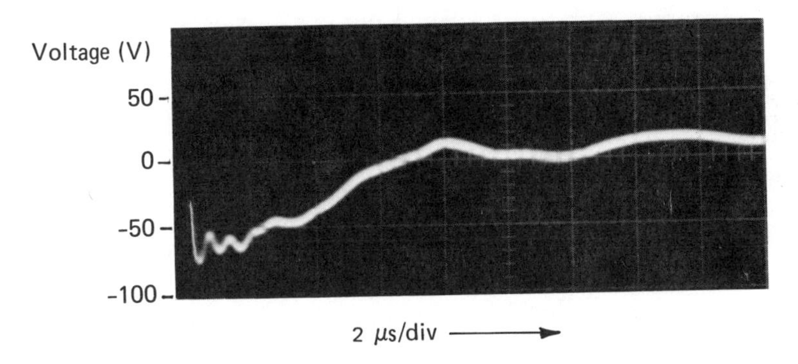

Fig 12
Same as Fig 11, but Further Expanded Time Scale to Show High-Frequency Ringing

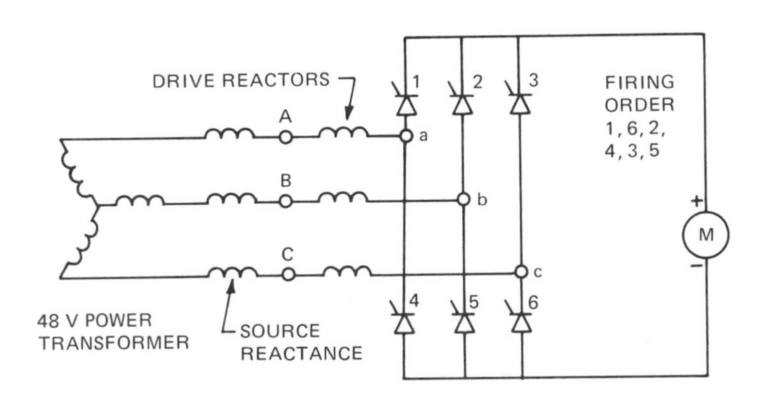

Fig 13
Three-Phase Full-Wave Rectifier Motor Control Power Circuits

The noise is produced when the current flowing in one line must suddenly stop and this same current must suddenly be established in another line. For example, assume that SCRs 1 and 6 have fired and are carrying the motor current, and now it is time for SCR 2 to fire and transfer the current from line A to line B. The lines have inductance preventing sudden changes in current, and so SCRs 1 and 2 are both conducting (short-circuited together) until this transfer is completed.

Figure 14 shows an oscillogram of the voltage *ab* going to zero as this transfer takes place. This phenomenon is called *line notching*. The notches are not as wide at reduced current, as shown in Fig 15. These noise prints were made on a 400 A line which had power factor correction capacitors on the power transformer secondary, resulting in a very small source impedance power system. For this reason the drive reactors were included as part of the source impedance to demonstrate line notching. In normal operation this 50 hp drive produces line disturbance on this power system, as shown in Fig 16.

For a further discussion of the effects of source reactance versus feeder and load reactance, load current, and firing angle, see [65].

Details of the line notch are shown in Figs 17 and 18. Current transfer details of the notch are illustrated in Fig 19.

4.1.2.3 Inductive Load

4.1.3 Circuit Switching

4.1.3.1 Multispeed Motors

4.1.3.2 Capacitor Banks

4.1.3.3 Transformer

4.1.3.4 Distribution Circuitry

4.1.3.5 Fluorescent Lights

4.1.4 Faults

4.1.4.1 Current-Limiting Fuses

4.1.4.2 Arcing Faults

4.1.4.3 Restrikes on Switching

4.1.5 Power-Frequency Electric and Magnetic Fields

4.1.5.1 Resistance Welding Machines

4.1.5.2 Power Lines

4.1.6 High-Frequency Narrowband Generators

4.1.6.1 Plastic Heating

4.1.6.2 Induction Heating

4.1.6.3 Diathermy

4.1.6.4 Arc Starters for Arc Welders

4.1.6.5 Radio Transmitters (Transceivers)

4.1.7 Radio-Frequency Broadband Generators. Most radio-frequency sources generate broadband noise. Of these, the characteristics of only a few have been well documented, for example, power transmission lines, automotive ignition, and radio-frequency-stabilized arc welders. Solid-state devices, for example, thyristors, are known to be great offenders in the genera-

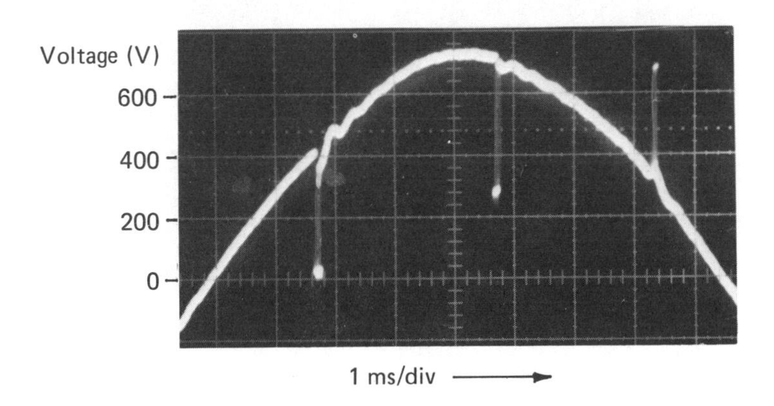

Fig 14
Typical Line Notching Transients Produced by Three-Phase Full-Wave Rectifier Motor Control
(Half-cycle of voltage *ab* in Fig 13 with motor loaded to produce 50 A rms in each ac line.)

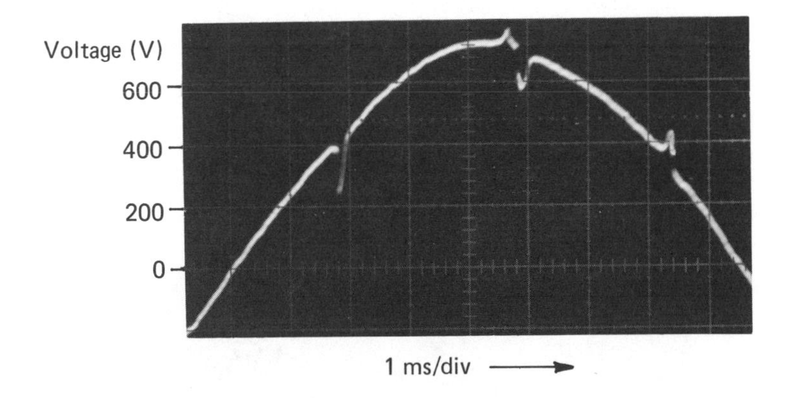

Fig 15
Voltage *ab* in Fig 13 with Motor Load Reduced to 15 A rms in Each AC Line

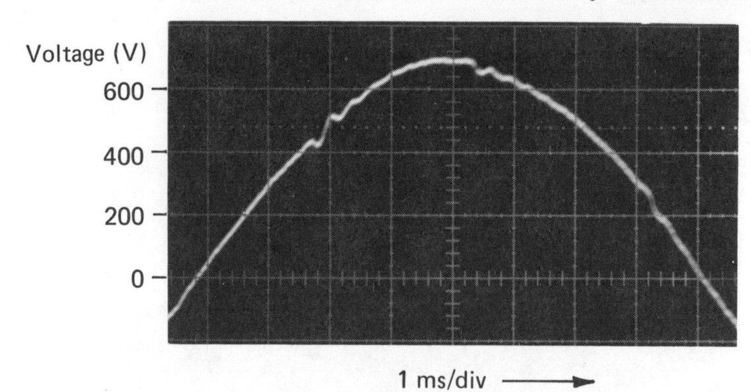

Fig 16
Voltage *ab* in Fig 13 with Motor Load to Produce 50 A rms in Each AC Line

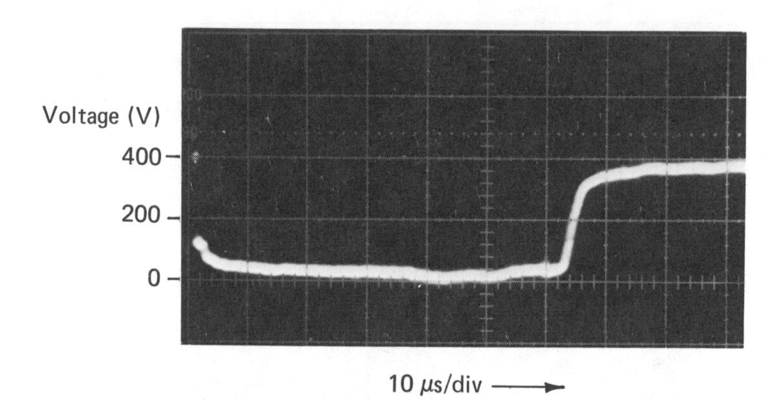

Fig 17
Same as Fig 14, but Expanded Time Scale to Show Notch Detail

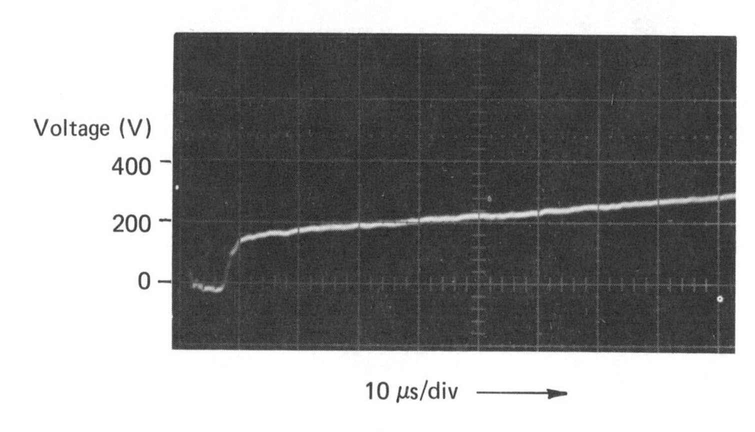

Fig 18
Same as Fig 15, but Expanded Time Scale to Show Notch Detail

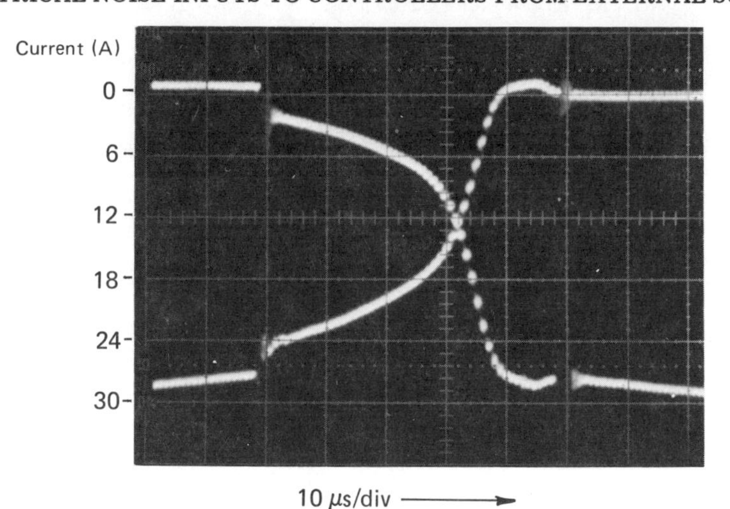

Fig 19
Current Transfer from Line *a* to Line *b* during Line Notch Shown in Fig 17

(Maximum d*i*/d*t* = 2.53 A/μs; initial d*i*/d*t* = 0.156 A/μs. Measurement was made with two 1200/5 Weston model 327 type 2 current transformers with 4 Ω resistance as secondary load.)

tion of radio-frequency noise. The radio-frequency characteristics of these devices have been under investigation.

The radio-frequency disturbance is transmitted to susceptible systems by conduction through power input lines and by radiation through space. Transmission of radio-frequency noise by radiation is complicated by the presence of environmental radio noise, which is composed of both atmospheric and man-made sources. The environmental radio noise varies from place to place as well as from time to time for the same place. Figure 20 shows the median values of radiated radio noise for various areas [3]. Table 1 is an example of the variation of radio noise with time [63].

An example of radiated radio noise from power lines, automotive traffic, and radio-frequency-stabilized arc welders is shown in Fig 21 [64]. It was concluded in [64] that below 25 MHz lower voltage transmission lines and radio-frequency-stabilized arc welders are the major incidental noise sources when the observer is within 100 ft of the source, and that above 40 MHz automotive traffic and lower voltage transmission lines are the major radio noise sources, with neither appearing to be consistently the greater when an observer is within 50 ft or less of the source.

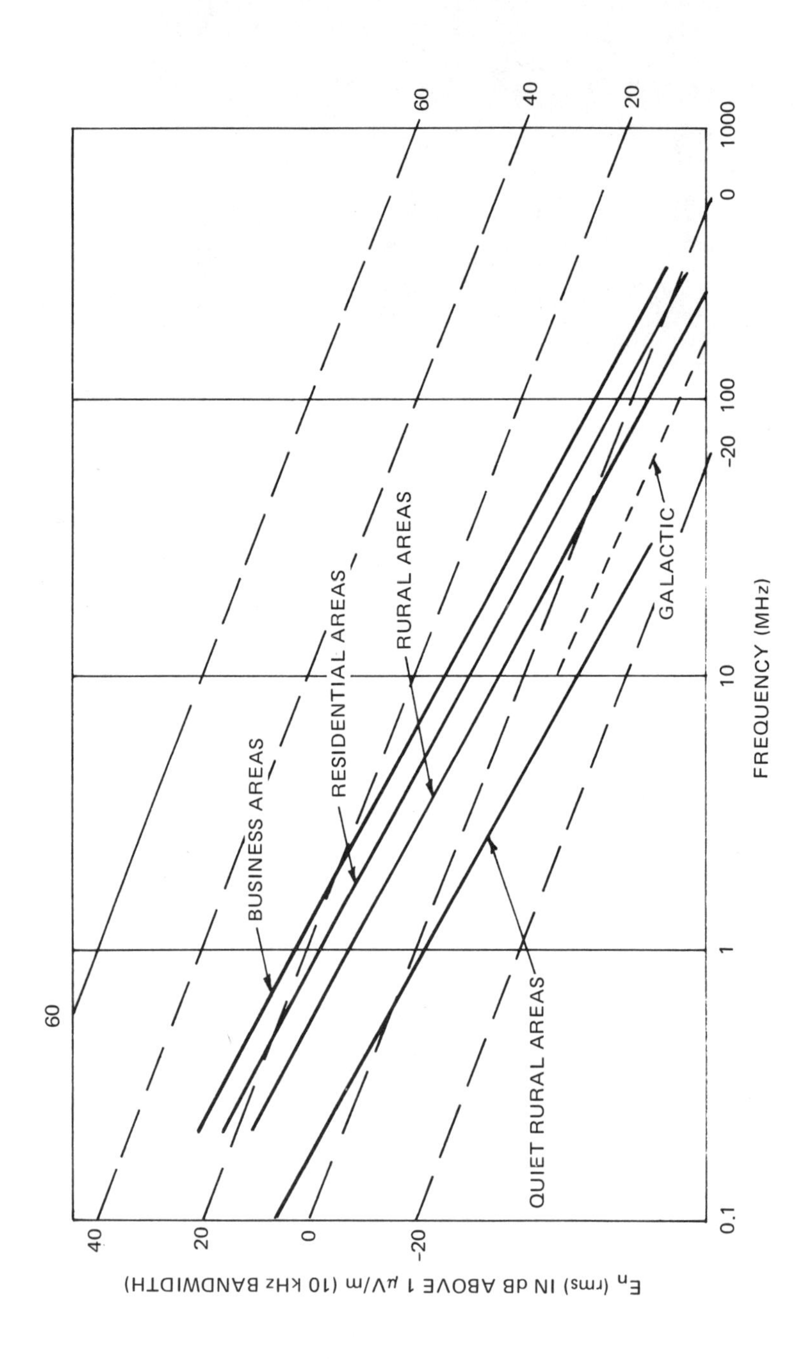

Fig 20
Median Values of Radio Noise [3]
(Omnidirectional antenna was near surface.)

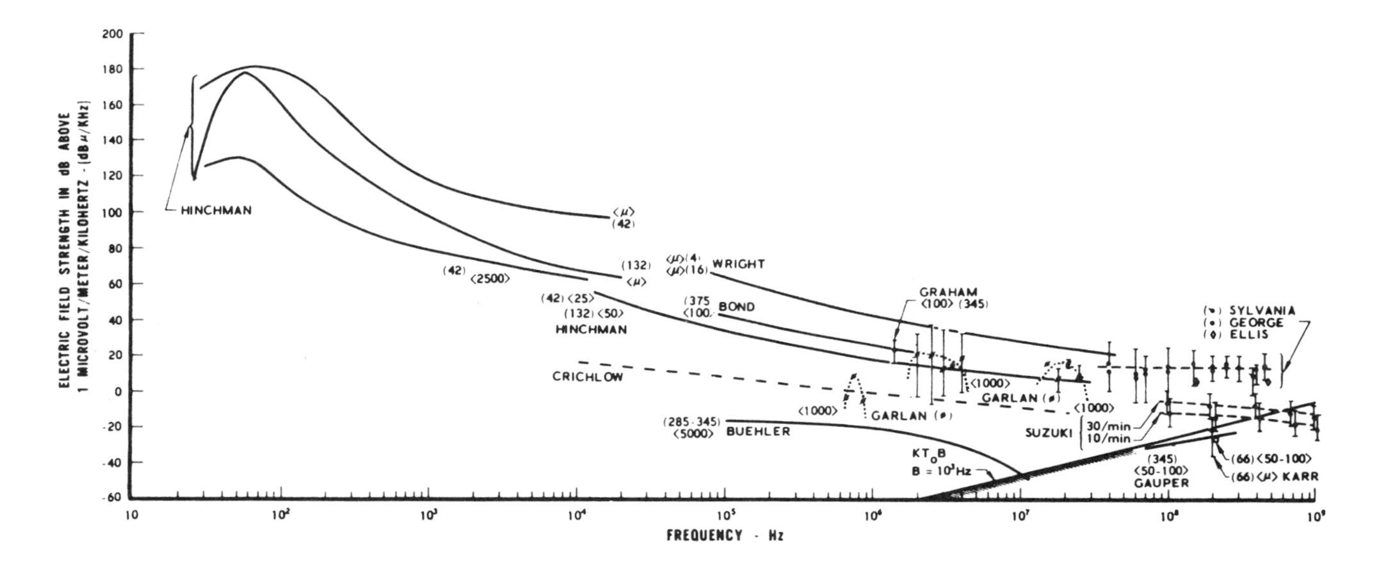

(Line voltages in kV—(40), (60), etc; observation point locations relative to sources in ft <μ> — <25>, <50>, <50-100>, etc; beneath conductor — power lines; • • • radio-frequency-stabilized arc welders at 1000 ft; - - - - - automotive ignition.)

Fig 21
Radiated Noise from Power Lines, Automotive Traffic, and Radio-Frequency-Stabilized Welders [64]

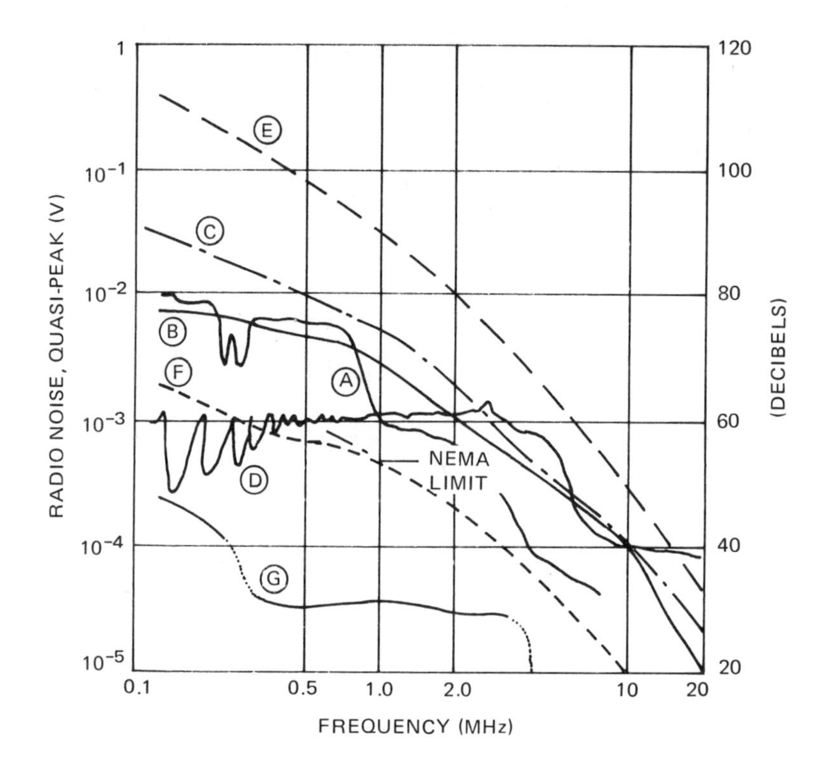

A 80 W, 30 kHz lighting inverter (experimental)
B Sonic cleaner, about 25 kHz
C Sonic cleaner, about 50 kHz
D Solid-state electric razor
E Dimmer, unsuppressed
F Dimmer, suppressed
G Controlled turn-on

Fig 22
Conducted Interference for *Power Conditioned* Equipment up to 100 W [35]

Figure 22 shows examples of conducted interference from solid-state-controlled consumer equipment [35]. Figure 23 shows an example of radiated radio-frequency noise from a thyristor-controlled dc propulsion system of a transit car [31], and Fig 24 is another example of radiated radio-frequency noise from a thyristor-controlled propulsion system [48].

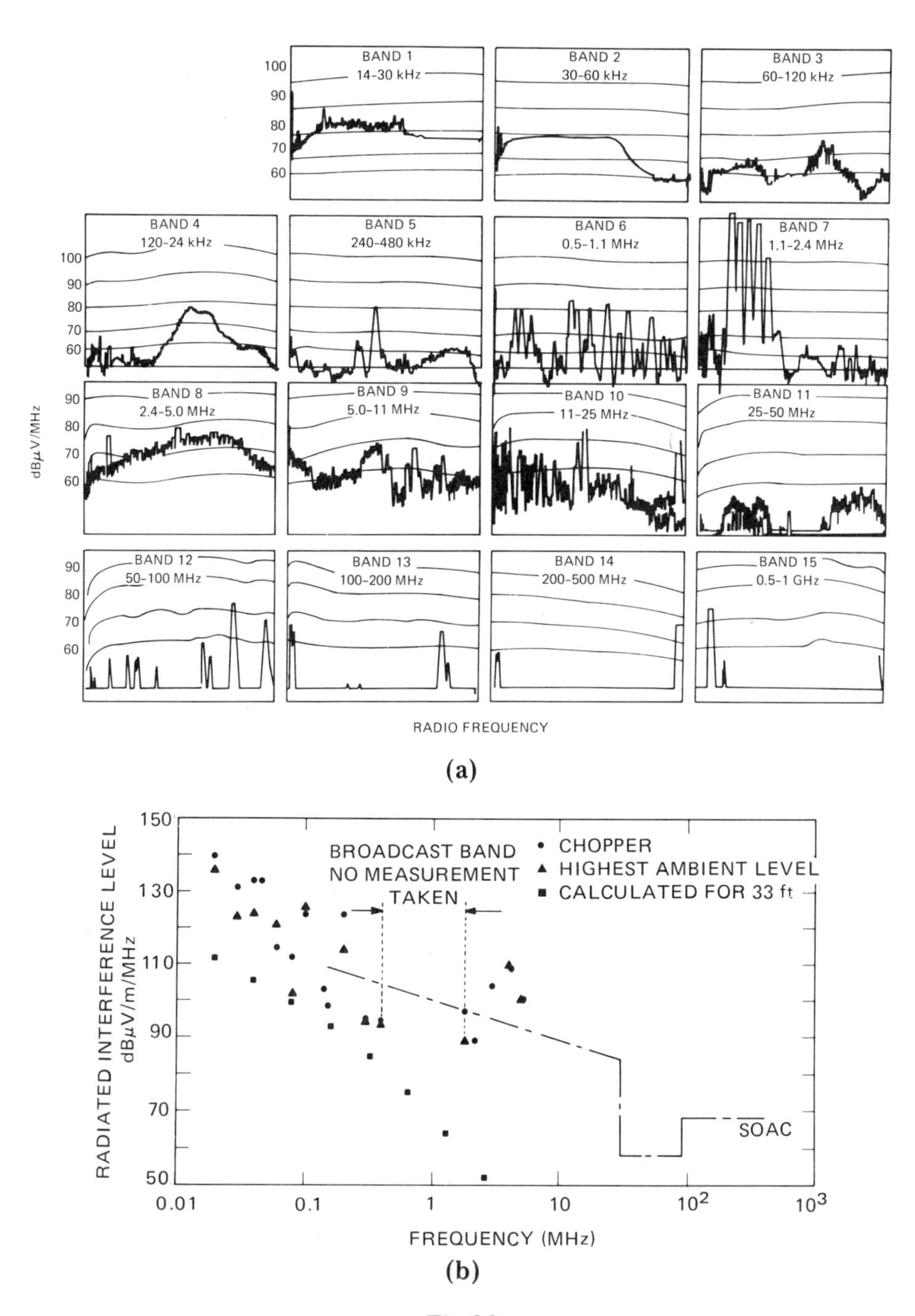

Fig 23
Radiated Peak Interference Levels from Chopper Car Running at 11 mi/h (17.6 km/h); Measurement Distance 33 ft (10 m) [31]. (a) As Measured. (b) Corrected for Antenna Factors

Table 1
Data Measured in Hybla Valley,
Fairfax County, Virginia,
on August 15 and August 25, 1955 [63]

Frequency (MHz/s)	Time	Signal Level (μV/m)	Signal Level (dBm/kHz)
236	Day	30	-115
	Night	1	—
273	Day	60	-108
	Night	1	—
291	Day	8	-129
	Night	1	—
350	Day	90	-107
	Night	1	—

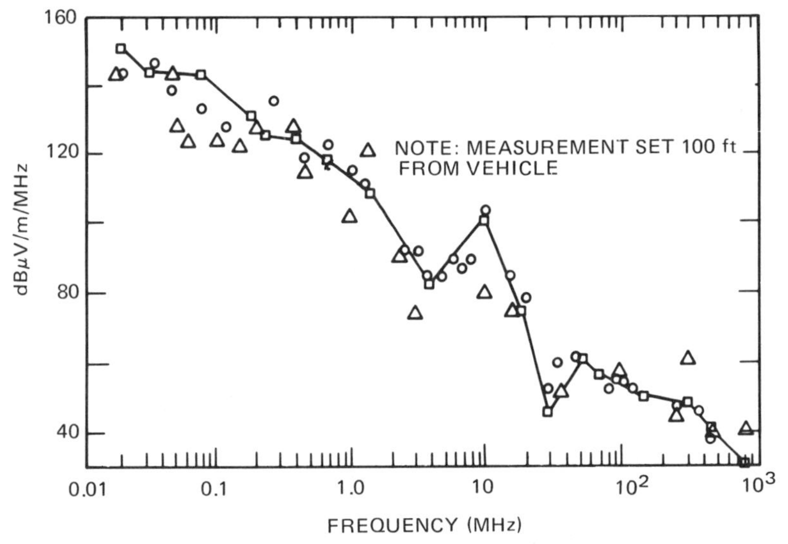

Fig 24
Radiated Peak Interference from AIRTRANS;
Measurement Distance 100 ft (30 m) [48]

4.1.8 Ferroresonance

4.1.9 Lightning and Electromagnetic Pulse

4.1.9.1 Lightning. An abundance of literature is available on lightning and protection against lightning of high-voltage apparatus of electrical power systems [22], [40], [41], [68]. Although research on the lightning protection of telephone cables started at Bell Laboratories in the 1940s [24] [66], interest in the effects of lightning on low-voltage systems in general started only about 20 years ago, when solid-state devices, such as diodes, transistors, and thyristors, were being used in control and communications systems.

The Telephone Association of Canada conducted a field investigation of lightning voltages induced in outside plant toll facilities at ten selected sites across Canada during the 1968 and 1969 lightning seasons [23]. Their results indicate that a standard test wave, with 1000 V peak and 10/1000 μs wave shape, simulates 99.8% of the lightning surges encountered in paired and coaxial cables, where 10 μs is the front time of the wave and 1000 μs is the decay time to half the peak voltage. Similarly, for open-wire circuits a more suitable test wave with 2000 V peak amplitude and 4/200 μs wave shape is required to simulate 99.8% of the lightning surges.

The General Electric Company performed transient-voltage measurements on two transit systems, the Chicago Transit Authority and the Long Island Railroad, during 1970–1972 [28]. The transient voltages were measured on the 600 V dc third-rail systems. Although switching surges and lightning surges were not differentiated, the measurements at the Chicago Transit Authority showed a significant increase in the number and magnitude of transient voltages during thunderstorm periods. Transient voltages exceeding 3000 V were attributed to lightning.

International standards efforts (for example, the International Electrotechnical Commission) favor the 1.2/50 μs wave shape as the standard for testing the transient-voltage withstand capability of low-voltage apparatus and systems. This wave shape has long been the standard wave shape for impulse testing of high-voltage apparatus.

4.1.9.2 Electromagnetic Pulse [25], [38], [43], [46], [49], [52], [53]. Radiation effects of nuclear explosions have long been recognized as harmful to humans, animals, and vegetation. However, electromagnetic pulses, although biologically harmless, can play havoc with communications, electronic, and electrical systems by inducing current and voltage surges through electrically conducting materials. In some cases the surges trip circuit breakers, shutting down a piece of equipment or power line. In other cases, individual components, such as semiconductor devices, or circuits are destroyed.

The electromagnetic pulse (EMP) has three transient effects. First, a pulse of ground current flows radially from the point of explosion. Second, a magnetic-field pulse propagates away from the point of explosion with the same vector components as those from a vertical dipole. Third, a corresponding pulse of electric field is propagated.

Results of extensive analyses on electromagnetic pulses are available. However, actual test results are classified. When considering the effects rather than the electromagnetic-pulse generation, one may replace the nuclear model with the more common phenomenon of a lighting stroke, which also produces the same three components of the electromagnetic-pulse effect. However, to make the lightning model simulate nuclear electromagnetic-pulse magnitudes equidistant from nuclear explosions in the megaton range would assume a lightning current far greater than has ever been measured. The electromagnetic field of lightning reaches its peak value in a few microseconds, while the electromagnetic field of an electromagnetic pulse rises to peak in a few nanoseconds.

4.1.10 Static Discharge

4.1.11 Cable Noise

4.1.12 Electrical Noise Measurement. Electrical noise is measured in both the time domain and the frequency domain. Time-domain measurements are usually performed on transient voltages and currents which are destructive to an apparatus or to a system. Frequency-domain measurements are generally performed on electric, magnetic, and electromagnetic fields which cause malfunction (or interference) on radio and television receivers and other electronic equipment.

Measurement of electrical noise is an extensive area, which is beyond the scope of this guide. However, for the sake of completeness, several references are cited for both time-domain [12], [6], [61] and frequency-domain [1], [2], [7], [13]–[15], [19] measurements.

4.2 Electrical Noise Susceptibility Classification

4.2.1 Introduction. Electrical noise is an unwanted voltage or current, or both, which appears in an electrical system. Commensurate with the characteristics of the system, electrical noise may be innocuous to the proper functioning of the system, or it may force the system to malfunction, or it may even lead to damage and destruction.

The electromagnetic compatibility (EMC) of an electrical system is its ability to perform its specified functions in the presence of electrical noise generated either internally or externally by other systems. In other words, the goal of electromagnetic compatibility is to minimize the influence of electrical noise.

Every electrical device or system has some immunity to electrical noise which is its inherent characteristic. If the strength of the oncoming electrical noise is below the threshold level of the receiver, be it a device or a system, special precaution is not necessary. However, if the strength of the electrical noise is above the threshold level, then techniques need to be applied to attenuate the electrical noise to a level below the threshold level of the receiver.

To be able to predict the degree of electromagnetic compatibility of a given system, one should know the following:

(1) Characteristics of the sources of electrical noise
(2) Means of transmission of electrical noise
(3) Characteristics of the susceptibility of the system
(4) Techniques to attenuate electrical noise

The purpose of this section is to define the criteria of susceptibility of electrical control systems to electrical noise. This discussion is based on [32].

4.2.2 System Boundaries. Electrical noise can reach the system through four different paths (Fig 25):

(1) Power feed lines
(2) Input signal lines
(3) Output signal lines
(4) Radiation

The susceptibility level of the system will be different for the four paths of entry of the electrical noise. Furthermore, even for the same path of entry, the susceptibility will depend upon the type of electrical noise. For instance, for the first three entry paths, the susceptibility will be different for common-mode noise from that for normal-mode noise. For radiated noise the susceptibility may depend upon the type of wave—electric, magnetic, or electromagnetic.

4.2.3 Malfunction Characteristics

4.2.3.1 Nondestructive Malfunction. The primary result of a nondestructive malfunction is malfunction of the system without necessarily leading to destruction of the control system, that is, the system stops malfunctioning and returns to normal operation after the electrical noise is removed. For instance, an electrical noise pulse injected between the base and the emitter of a transitor may inadvertently switch the transistor on or off. Similarly, an electrical noise between the gate and the cathode of a thyristor may turn the thyristor on; or a forward voltage of high dv/dt may switch the thyristor on without necessarily damaging it. As another example, an electrical noise coupled through the power supply into the amplifier stage of an analog regu-

Fig 25
System Boundaries for Penetration of Electrical Noise

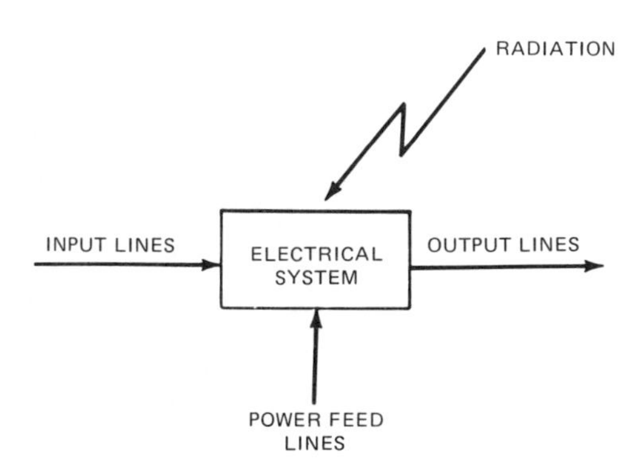

lator may cause a disturbance to the controlled parameter, without any actual damage to the regulator itself.

4.2.3.2 Destructive Malfunction. A destructive malfunction causes damage to the system irreversibly, that is, the system continues to malfunction even after removal of the destructive electrical noise, until the damaged components are replaced. For instance, a reverse voltage of sufficient magnitude across the base-emitter junction of a transistor can lead to destruction of the device. Similarly, a still higher voltage across the anode and the cathode of the thyristor would lead to irreversible breakdown of the device.

4.2.3.3 Nondestructive versus Destructive Malfunction. It may seem that once a system is protected against the nondestructive malfunction, it would also be protected automatically from the destructive malfunction. This is not always true. The characteristics of the electrical noise sources for these two types of malfunctions are generally quite different from one another. Furthermore, the suppression techniques may be quite different for these two types of electrical noise sources.

As an illustration, consider a pair of control lines running parallel to a power line. The pair of control lines are twisted to minimize pickup caused by the 60 Hz magnetic field of the power line. However, high-voltage transients may appear on the power line caused by power switching during faults or lightning. The induced voltage across the control lines may be high enough to cause a breakdown. The destructive electrical noise may be attenuated even below the nondestructive threshold level by elaborate shielding and grounding. However, this will be expensive. Alternatively, the destructive electrical noise may be attenuated below the destructive level by inexpensive surge suppression techniques, thereby letting the system temporarily malfunction for the duration of the transient.

As a second illustration, a conducted nondestructive electrical noise may be attenuated by a filter at the input of the equipment so that no malfunction occurs. The filter size will be larger, and hence the filter more expensive, if it is specified to make a destructive electrical noise innocuous to the system. The combination of a surge suppressor and a filter will be optimal in this case.

4.2.4 Electrical Noise Characteristics. Electrical noise can, in general, be divided into two classes: continuous wave and transient.

4.2.4.1 Continuous-Wave Electrical Noise. A continuous-wave electrical noise can further be subdivided into narrowband and broadband.

A narrowband continuous-wave electrical noise is, ideally, a disturbance at a single frequency with zero bandwidth. The single-frequency disturbance from an oscillator or a transmitter is a narrowband continuous-wave electrical noise.

A broadband continuous-wave electrical noise is a disturbance which contains energy covering a wide frequency range. A repetitive rectangular pulse is a broadband electrical noise (Fig 26).

A broadband electrical noise can be mathematically expressed, by Fourier

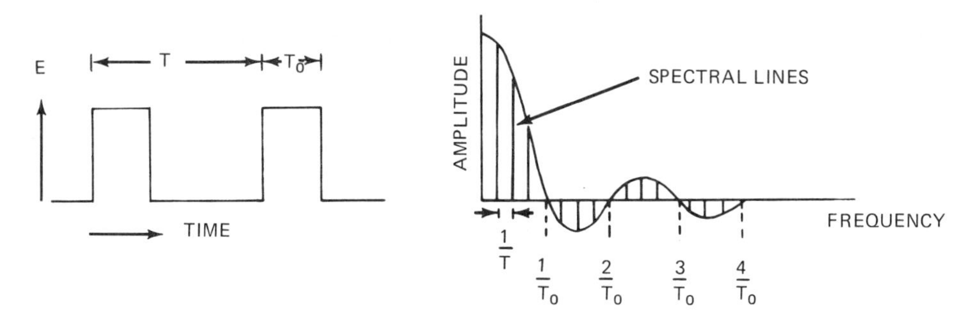

Fig 26
Spectrum of Rectangular Pulse Train

series, as the summation of narrowband continuous-wave functions whose amplitudes are frequency dependent. For instance, the repetitive rectangular pulse of Fig 26 can be expressed as

$$F(t) = \frac{ET_0}{T} + \sum_{n=1}^{\infty} \frac{2}{n}\frac{E}{\pi} \sin\left(\frac{n\omega T_0}{2}\right) \cos(n\omega t)$$

$$= \frac{ET_0}{T} + \sum_{n=1}^{\infty} S(\omega, n) \cos(n\omega t)$$

where

$\omega = 2\pi/T$.

The plot of $S(\omega, n)$ as a function of frequency f is called the spectrum of the function $F(t)$ in the frequency domain.

4.2.4.2 Transient Electrical Noise. In many cases, electrical transients are the main cause of control system malfunctions, both destructive and non-destructive. Most transients are broadband and occur at random. Of the four transients shown in Fig 27, the first three are broadband and the last one is narrowband.

Figure 27(a) illustrates a rectangular pulse which is sometimes used as an approximation of actual transients for the sake of simplicity in analytical work. Figure 27(b) is a standard transient voltage specified in military specifications [19]—[21]. The duration of the first loop is 10 μs. The front time is not specified. Figure 27(c) shows an industry standard transient voltage [5]. It has a front time of 1.2 μs, a time to half-value of 50 μs, and is designated a 1.2/50 μs wave.

This wave shape is an approximation of the transient voltage generated by a lightning discharge on a power transmission line or a distribution line. The basic impulse insulation level (BIL) of high-voltage apparatus is specified by

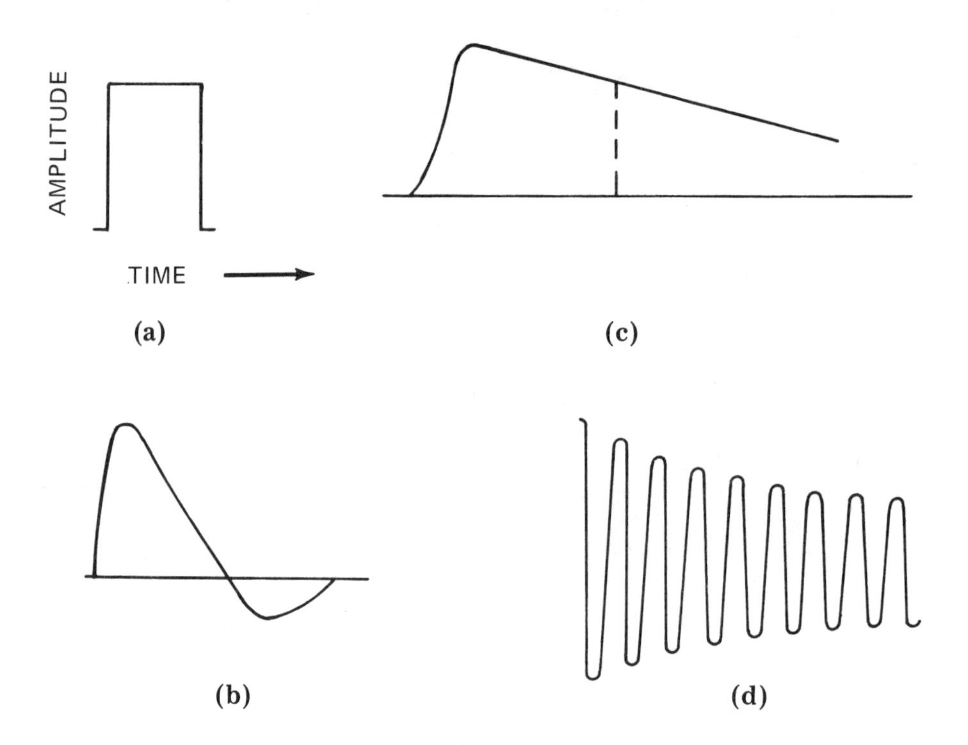

Fig 27
Illustration of Transient Electrical Noise

this wave shape, the actual magnitude being dependent on the voltage class of the apparatus. Figure 27(d) has been proposed as the surge withstand capability (SWC) test for solid-state relays [4]. This is an oscillatory wave of 1.5 MHz nominal frequency, with the envelope decaying to 50% of crest value in not less than 6 μs, and a pulse repetition rate of 50 pulses per second for a period of not less than 2 μs.

Transients can be represented in the frequency domain by Fourier integrals for analytical purposes. For instance, a rectangular pulse can be represented in the frequency domain, as in Fig 28, by

$$S(\omega) = \left(\frac{2}{\omega}\right) \sin \left(\frac{\omega T}{2}\right)$$

However, electrical transients are almost universally represented in the time domain because they are convenient to generate and measure in the time domain.

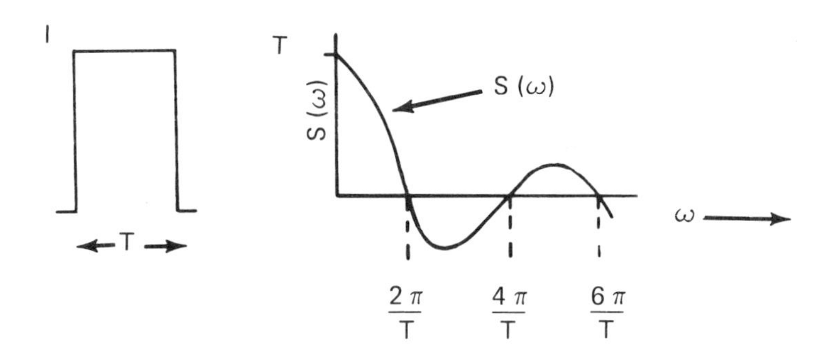

Fig 28
Spectral Density of Rectangular Pulse

4.2.5 Criteria of Susceptibility Measurement. It is known that the susceptibility of the most commonly used semiconductor devices (diodes, transistors, thyristors, integrated circuits) to electrical noise depends largely upon the magnitude of the voltage or current of the impressed electrical noise and its charge or energy content, or both [26], [27], [29], [37], [55], [71]. On the other hand, the characteristics of passive electrical circuits are primarily frequency dependent. The susceptibility of an analog control circuit is a function of the specified accuracy of the circuit, that is, of the deviation or output tolerance of the circuit without any external electrical noise present. As long as the control circuit output stays within this normal tolerance in the presence of an electrical noise, the circuit is considered to be not susceptible to the type and level of electrical noise. It is obvious then that an electrical system containing both the passive resistance-inductance-capacitance (*RLC*) elements and the active devices would have very complex susceptibility to electrical noise.

The rational approach appears to be to characterize system susceptibility in accordance with the characteristics of the impressed electrical noise, that is, continuous wave and transient.

4.2.5.1 Susceptibility to Continuous-Wave Electrical Noise. The conducted continuous wave susceptibility is generally measured by injecting a voltage of a certain frequency to the input of the receiver (or system) and determining the voltage level at which the receiver malfunctions [19]—[21]. This procedure is repeated over a specified range of frequencies. This method does not appear to be completely justified. The susceptibility of many systems to electrical noise depends upon the interaction between the impedances of the electrical noise source and the system. The lower the impedance of the system and the higher the impedance of the source, the lower will be the susceptibility of the system to an external electrical noise. This difference is not recognized in the present method of determining susceptibility. This is illustrated by an example. Let us consider two identical systems which have a conducted narrowband susceptibility of x V at a specified frequency. Let a capacitor be connected across one of the systems to improve its susceptibility. However, according to the present method of testing, both systems will still have the same susceptibility because the present method requires that equal voltages be impressed across the terminals of each of the two systems. In actuality, however, a lower voltage will appear across the input terminals of the system that has the capacitor connected across its input terminals for the same source of electrical noise.

This discrepancy may be corrected to some extent by standardizing both the source (signal generator) impedance and the system impedance, and reducing the required susceptibility limit in proportion to the actual system impedance. This procedure has the drawback that the source and system impedances are generally complex, having both active and reactive parts which are frequency dependent.

An alternate method is to specify the conducted narrowband continuous-wave susceptibility of a system in terms of volts and volt-amperes as a function of frequency. In other words, at each specified frequency of the electrical noise source, the susceptibility of the system should be measured in terms of the impressed voltage across the input terminals of the system and the voltamperes injected by the electrical noise source into the system. The requirements of volts and voltamperes do not have to be met simultaneously. To illustrate, let us consider again the example of the two systems, one having an input capacitor, but which are otherwise identical. To determine the susceptibility of the system with the input capacitor, the output voltage of the electrical noise source at a specified frequency is gradually increased, while the voltage across the input of the system as well as the injected voltamperes are monitored. If the input capacitance is high enough, then the limit of the input voltamperes is reached. For the second system, depending upon the input impedance, the required input voltage magnitude may be reached before the limit of the voltampere requirement. In this way the designer of the system will be free to reduce the system susceptibility either by increasing the threshold or by decreasing the system impedance, or both. The electrical noise source (signal generator) is also free from any constraints, except that it should have the capability to inject the required volts and voltamperes.

4.2.5.2 Susceptibility to Transient Electrical Noise. The susceptibility to transient electrical noise is another matter. The continuous-wave susceptibility cannot cover this area because of the limited energy contained in transient electrical noise. Moreover, the wave shape of transient electrical noise is a significant parameter because the performance of many modern semiconductor devices depends upon the di/dt and dv/dt of the electrical noise. Therefore it is proposed that the susceptibility to transient electrical noise be

specified by the following parameters:
(1) Peak of transient voltage
(2) Wave shape of voltage (front time and duration)
(3) Energy

Similar to the case of continuous-wave susceptibility, it may not be possible (or necessary) to satisfy all the above three requirements simultaneously. The susceptibility to transient electrical noise will be governed by the designed threshold level and impedance level of the system. Many control systems contain semiconductor devices which are sensitive to di/dt and the dv/dt of the impressed source. Special networks are designed, either internally or externally to these systems, to minimize the dv/dt effects. Therefore insistence on the specified voltage wave shape across the input terminals of the system would be either to face an impossible situation or to disregard the merits of the special design features of the system.

The specifications concerning the transient electrical noise susceptibility of communications/control systems are relatively new. These are mostly required by military standards [19]—[20]. However, it is believed that these specifications are not complete. For instance, none of these specifications has any limit on the front time of the transient electrical noise. Moreoever, the energy of the applied electrical noise is not explicitly specified. These standards recommend calibration of the transient-voltage generator by connecting a fixed resistor, for example, 5 Ω, across its terminals. Although this implicitly specifies the energy-handling capability of the electrical noise source, the actual energy injected into the system is not known when the fixed resistor is replaced by the system under test, and the output voltage of the electrical noise source is readjusted to the specified value. The susceptibility of the system, as measured by this method, may depend considerably on the internal design of the transient-voltage generator.

A method is proposed which will measure the susceptibility of the system to transient electrical noise without any regard to the internal design of the transient-voltage generator. It is proposed that the generator be set up to produce the specified wave shape on open circuit. The generator is then connected to the system. The voltage across the system terminals as well as the energy injected into the system are simultaneously measured. The peak terminal voltage is adjusted until either the specified peak voltage is reached or the specified energy is injected into the system, whichever occurs first.

Transients in electrical systems occur at random. Two transients are seldom alike, either in wave shape or in amplitude. However, industry-standard wave shapes are already in existence which can be applied to test the susceptibility of equipment to transient electrical noise.

Transient voltages may last for microseconds as well as milliseconds. For miscrosecond transients a standard wave shape of 1.2/50 μs is as shown in Fig 27(c) [5]. This wave shape in the *chopped* mode is represented by the vertical dashed line in Fig 27(c). Chopping denotes a flashover of an insulating path, such as a pair of separating contacts. Showering arc, as discussed in [67], can be represented by a thyratron or a thyristor, and the time to chop can be varied to produce the worst case by logic circuitry.

In [4] a damped oscillatory sine wave has been proposed, as shown in Fig 27(d). It has a frequency range of 1.0—1.5 MHz, the envelope decaying to 50% of the first crest in not less than 6.0 μs.

For representing millisecond transients, one half-cycle of the 60 Hz wave can be used. This type of transient-voltage wave shape has been successfully used to test semiconductor devices.

It is admitted that the proposed transient-voltage wave shapes will not represent all possible types of transient voltages which may be generated in the various electrical systems. As basically there are two types of transients, short duration and long duration, the proposed wave shapes would simulate a broad range of transients. Under special circumstances, other wave shapes should also be considered.

4.2.5.3 Overall Susceptibility. The susceptibility of a system will be different for the various means of entry (Fig 25) of the same type of electrical noise. This is not only due to the characteristics of the active devices, but it is also true because the incoming electrical noise encounters different impedance networks. The susceptibility of a system also depends upon the mode of propagation of the electrical noise, namely, common mode or normal mode. The susceptibility to electrical noise can be categorized as shown in Fig 29.

4.2.6 Susceptibility of Control Systems. The classification in Fig 29 may appear to be formidable at first glance. However, it can be considerably simplified. Almost all destructive faults are caused by transient voltages. Therefore the susceptibility criterion of destructive faults caused by continuous waves may be eliminated.

The withstand capability of a system to common-mode electrical noise is generally much higher than that to normal-mode electrical noise, and the ratio of the two withstand capabilities can be expressed as a number A, as shown in Table 2.

The ratio of output signals to input signals is the gain of the system. It is logical to express the ratio of these withstand capabilities also in terms of the gain β, shown in Table 2. Similarly, the ratio of the withstand capabilities of the input lines to the power feed can be expressed as equal to the ratio of the highest input signal voltage level to the system voltage α, shown in Table 2.

Therefore, the only parameter to be specified is the withstand capability of the power feed lines; α, β, and A will be known from the requirements of the system.

An example is shown in Table 2. For a particular system the normal-mode withstand capability of the power feed lines is specified as x V for temporary faults. Once this is specified, the others will follow suit as shown in

Table 2
Transient Susceptibility Requirements
(Temporary Fault)

Power Feed Lines		Input Lines		Output Lines	
Normal Mode	Common Mode	Normal Mode	Common Mode	Normal Mode	Common Mode
x (V)	Ax (V)	ax (V)	Aax (V)	βax (V)	$A\beta ax$ (V)
y (J)	A^2y (J)	a^2x (J)	A^2a^2x (J)	β^2a^2x (J)	$A^2\beta^2a^2x$ (J)

a = (highest input signal level)/(system voltage).
β = system voltage gain.
A = (withstand capability to common-mode noise)/(withstand capability to normal-mode noise).

Fig 29
Classification of Susceptibility to Electrical Noise

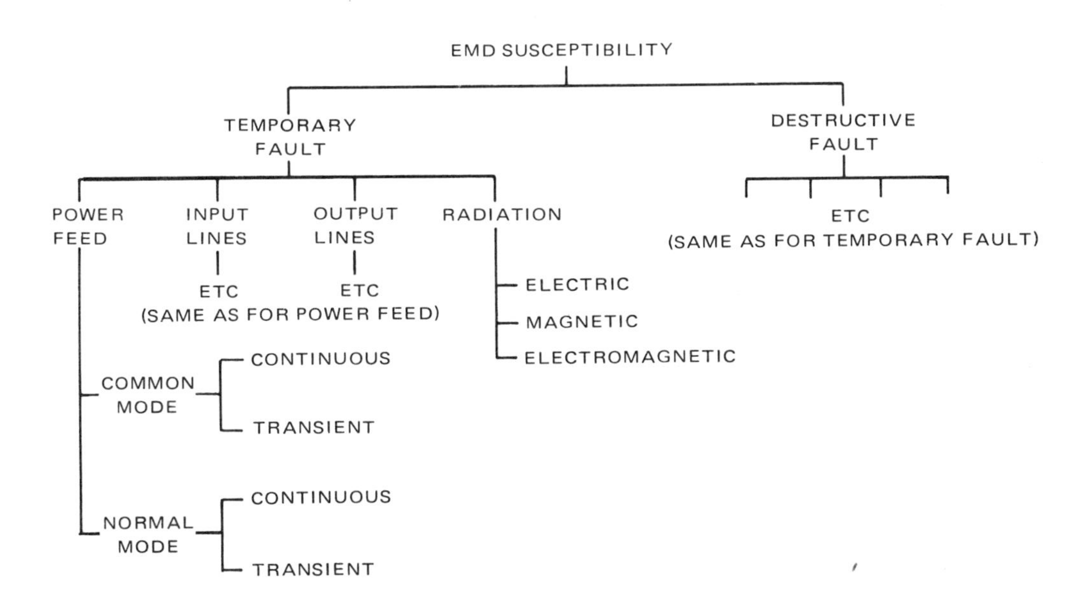

Table 2. Once the withstand capabilities for temporary faults are thus specified, the corresponding withstand capabilities for destructive faults can be obtained by multiplying by a factor to be determined by the system requirements.

Once the transient-voltage withstand capabilities are specified the transient-energy withstand capabilities can be similarly specified, starting with the normal-mode power feed withstand capability. For the transient-energy capabilities the multiplying factors are squared because energy is proportional to the voltage squared.

The same procedure should be used for both the short-duration and the long-duration transients.

The specifications on the continuous-wave withstand capabilities are similar. However, in this case instead of specifying a number x, a curve of voltages versus frequency has to be specified. The other steps are similar.

The numbers x and y and the voltage versus frequency curves will be different for different systems. It is recognized that this is a difficult task. It will be completed only when the characteristics of electrical noise in various systems can be estimated, and suppression to a known level can be obtained *economically*. Much work needs to be done in this area.

4.2.7 Methods of Measurement [30]

4.2.7.1 Measurement of Transient Susceptibility. Three apparatus are needed for the measurement of transient susceptibility:

(1) Transient-voltage generator
(2) Apparatus to measure transient voltage
(3) Apparatus to measure transient energy

4.2.7.1.1. Generation of Transient Voltage. The schematic of a transient voltage generator for generating the 1.2/50 μs voltage wave is shown in Fig 30. The wave shape is determined by the parameters C_1, C_2, R_1, and

Fig 30
1.2 / 50 μs Wave Transient-Voltage Generator

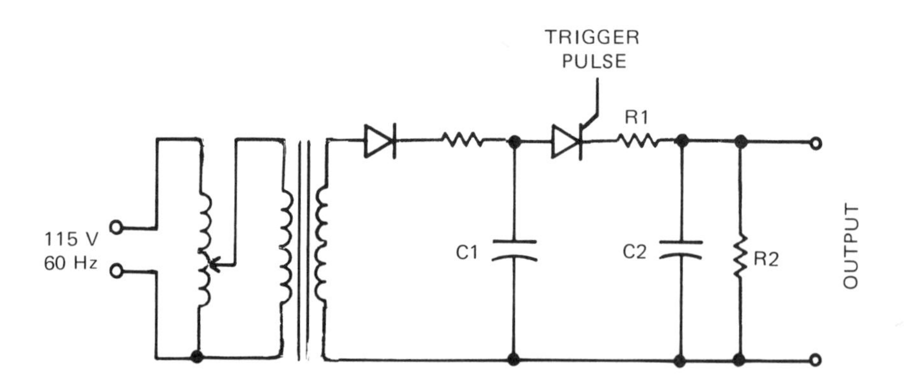

R_2. A fairly good estimate of the peak magnitude and the wave shape of the open-circuit output voltage can be made from the following equations:

$$\text{peak of output voltage} = V_1 \left(\frac{R_2}{R_1 + R_2}\right) \left(\frac{C_1}{C_1 + C_2}\right)$$

where

$$\text{time to crest} = 3R_1C_2$$
$$\text{time to half-value} = 0.8R_2C_1$$
V_1 = charging voltage of capacitor Cl.

The transient voltage [Fig 27(d)] is an oscillatory wave having a frequency range of 1.0-1.5 MHz, a voltage range of 2.5-3.0 kV crest of the first half-cycle peak, and an envelope decaying to 50% of the crest value of the first peak in not less than 6.0 μs from the start of the wave. The source impedance of the surge generator used to produce the test wave shall be 150 Ω. The test wave is to be applied to a test specimen at a repetition rate of not less than 50 tests per second for a period of not less than 2.0 s [4].

The test is a design test for relaying systems, in particular static relaying systems. The schematic of a typical test wave generator is shown in Fig 31. For its adaptation as a test voltage to determine the susceptibility levels of the various control systems, its magnitude has to be modified to suit the specific requirement.

The schematic of the 8 ms transient-voltage generator is shown in Fig 32. The open-circuit output voltage is given by the equation

$$V_0 = V_i \left(\frac{C_1}{C_1 + C_2}\right) \left[1 - \cos\left(\sqrt{\frac{t}{LC}}\right)\right]$$

where V_0 is the open-circuit output voltage, V_i is the charging voltage of capacitor C1, and $C = C_1C_2/(C_1 + C_2)$.

Fig 31
Oscillatory Wave Transient-Voltage Generator

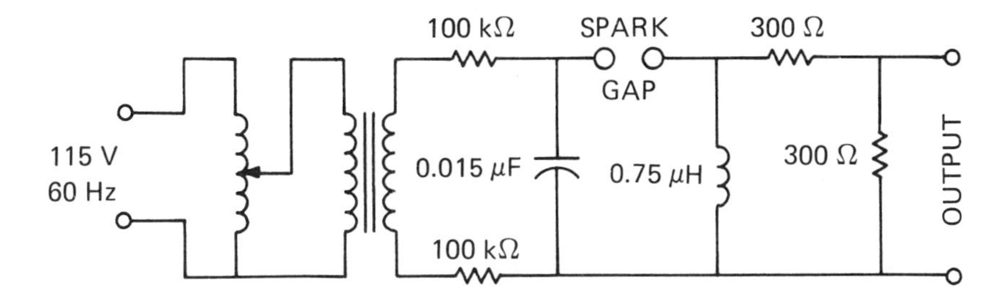

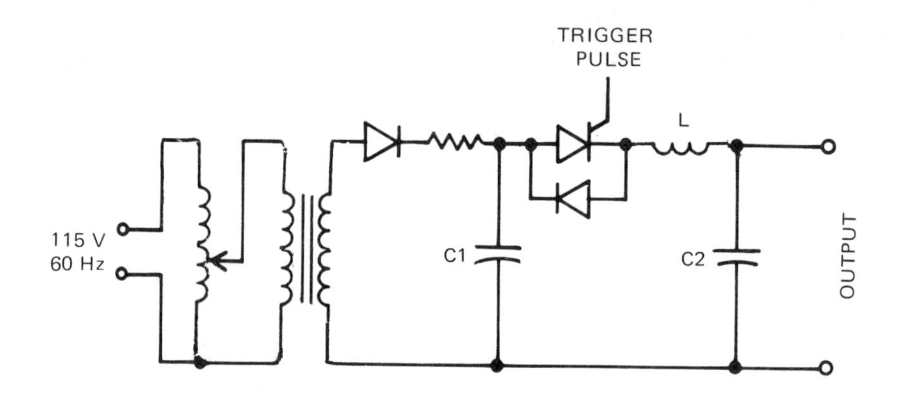

Fig 32
8 ms Wave Transient-Voltage Generator

4.2.7.1.2 Measurement of Transient Voltages. Transient voltages are generally measured by a cathode-ray oscilloscope which records both the wave shape and the magnitude of the transient. When the information on wave shape is not required, memory voltmeters may be used to record the peak magnitude of the transients.

4.2.7.1.3 Measurement of Transient Energy. The total energy during the transient condition is given by

$$E_T = \int_{t_1}^{t_2} (V_s + V_t)(i_s + i_t)\, dt$$

$$= \int_{t_1}^{t_2} V_s i_s\, dt + \int_{t_1}^{t_2} V_s i_t\, dt + \int_{t_1}^{t_2} V_t i_s\, dt + \int_{t_1}^{t_2} V_t i_t\, dt$$

The transient energy is given by

$$E_t = E_T - \text{steady-state energy during transient}$$

$$= E_T - V_s i_s\, dt$$

$$= \int_{t_1}^{t_2} V_s i_t\, dt + \int_{t_1}^{t_2} V_t i_s\, dt + \int_{t_1}^{t_2} V_t i_t\, dt$$

where

V_s = steady-state voltage
V_t = transient voltage
i_s = steady-state current
i_t = transient current

The difficulty of integrating three separate terms and summing may be considerable. A practical method would be to measure the total energy during the transient and subtract the steady-state energy from it during the same period of time. The steady-state energy is easily calculable. The total energy during the transient can be determined by any of the following methods:

(1) The total voltage ($V_s + V_t$) and the total current ($i_s + i_t$) are simultaneously displayed and photographed on a dual-beam cathode-ray oscilloscope. The magnitudes of the total voltage and current at various instants are multiplied, and the resulting instantaneous total voltamperes is plotted against time. The area under the voltampere versus time curve is measured to calculate the total energy in joules.

(2) The instantaneous total voltage and current are multiplied by an electronic multiplier, and the instantaneous total voltamperes is displayed and photographed on the cathode-ray oscilloscope. The area under this curve will be the total energy.

(3) The instantaneous total voltage and current are electronically multiplied, integrated with time, and measured on a calibrated indicating instrument directly in joules.

Although the labor involved in the first method does not become any lighter if the steady-state system is direct current, the procedures for the second and third methods become easier for a dc system. For the second method when the output of the multiplier is fed to the cathode-ray oscilloscope through a blocking capacitor, the steady-state voltamperes is automatically subtracted from the total voltamperes. Similarly for the third method, if the output of the multiplier is fed to the integrator through a blocking capacitor, the energy contained in the transient can be read directly.

The schematic of the susceptibility test is shown in Fig 33. This is only applicable when a transient voltage of positive polarity is applied to the positive bus of the system. To isolate the steady-state dc power from the transient-voltage generator, a decoupling capacitor is necessary at the output of the transient-voltage generator. The dc power source needs also to be isolated from the transient-voltage generator. A string of diodes on the positive terminal of the dc power source can be used for such isolation. To measure the transient energy injected into the equipment under test B, the voltage across and the current through B is fed into the multiplier F, and the output of F is in turn fed into an integrator. The output of the integrator is displayed on the cathode-ray oscilloscope G.

An isolating network, consisting of diodes and inductor, is used for the negative-polarity transient voltage (Fig 34). The test setup is otherwise similar to that for the positive-polarity transient voltage.

4.2.7.2 Measurement of Continuous-Wave Susceptibility. The measurement of the susceptibility to continuous-wave disturbances is similar to that of the transient susceptibility, except that it is required to measure the

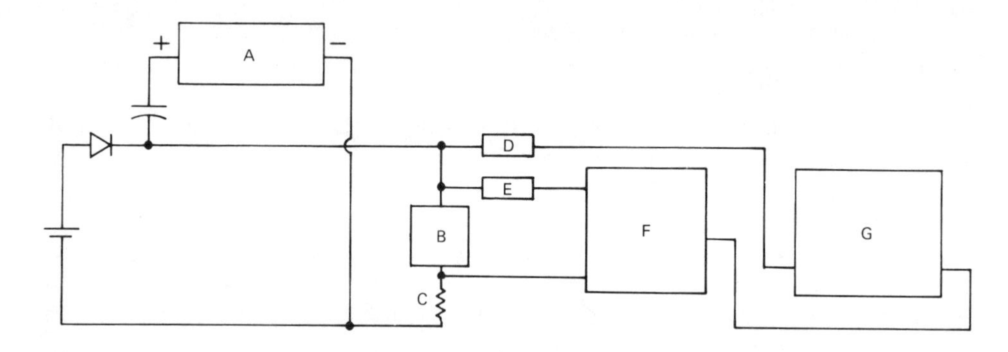

A Transient-voltage generator
B Equipment under test
C Coaxial shunt
D,E Voltage probes
F Multiplier
G Dual-beam cathode-ray oscilloscope

Fig 33
Transient Susceptibility Test, Positive Wave

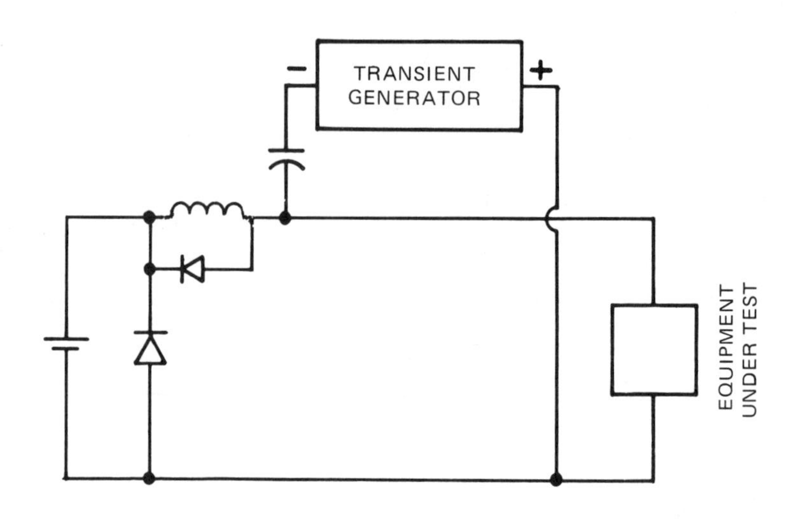

Fig 34
Transient Susceptibility Test, Negative Wave

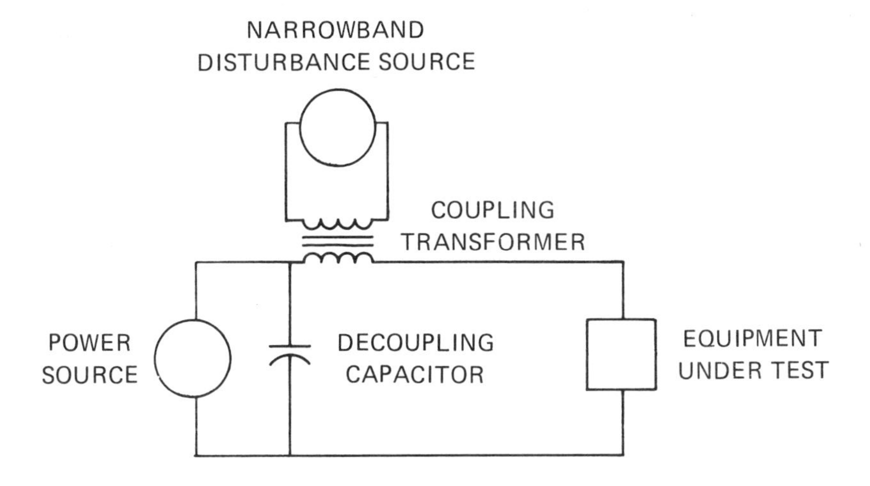

Fig 35
Test for Susceptibility to Narrowband Disturbances

voltamperes of the disturbance instead of the energy. Figure 35 shows a schematic of the experimental setup.

The continuous-wave disturbance source and the coupling transformer are commercially available. The decoupling capacitor is used to bypass the continuous-wave disturbance from the power source. To measure the voltamperes of the disturbance, the output of the multiplier, as shown in Fig 33, is connected directly to the cathode-ray oscilloscope instead of through the integrator.

4.2.7.3 Measurement of Radiated Susceptibility. The susceptibility of electrical equipment to radiated noise may be even more critical than its conducted susceptibility. Radiated susceptibility tests are generally performed by placing the test object very near the radiating antenna (at 1 m) in a shielded measuring room. A true radiation field does not develop at such a short distance within the shield enclosure. What is measured is the susceptibility to the induction field, that is, the susceptibility to the electric field and to the magnetic field.

4.2.7.3.1 Electric-Field Susceptibility Test [19]. The susceptibility to the radiated electric field is generally measured in the frequency range of 14 kHz to 1 GHz. The following apparatus will be required for this test:

(1) Signal (noise) source
(2) EMI meter
(3) Antennas
(4) Output monitor to check performance of test sample

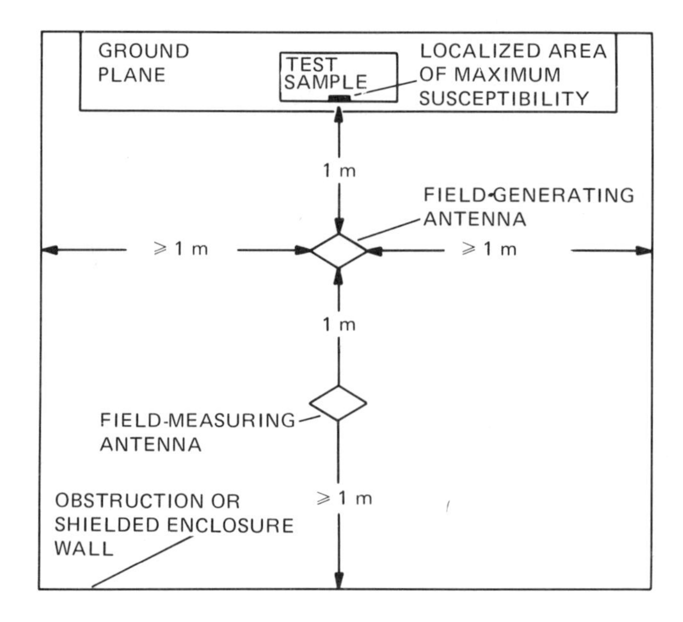

Fig 36
Antenna Placement for Radiated Susceptibility Measurements

No point of the field-generating and field-measuring antennas shall be less than 1 m from the walls of the enclosure or obstruction (Fig 36). In the frequency range of 14 kHz to 25 MHz, a 41 inch rod antenna (electrical length = 0.5 m) and an appropriate matching network with a square counterpoise shall be used. A biconical antenna is used in the frequency range of 20 MHz to 200 MHz, and a log-conical antenna in the 200 MHz to 1 GHz range. Generally three frequencies per octave are selected. The output of the signal generator is adjusted so that the generated fields at the test sample correspond to the applicable limit. The specified field strengths are measured by placing a field-measuring antenna at the same distance or relative location where the test sample will be placed. Generally an electric field of 1 V/m is considered adequate for measuring the susceptibility to the radiated electric field. The equipment under test shall perform within its specified limits when subjected to this 1 V/m electric field.

Where large fields (100 V/m) are required, the conventional antennas cannot be used because they are power limited and inefficient. In such special cases, parallel-plate lines and their variations are recommended [72].

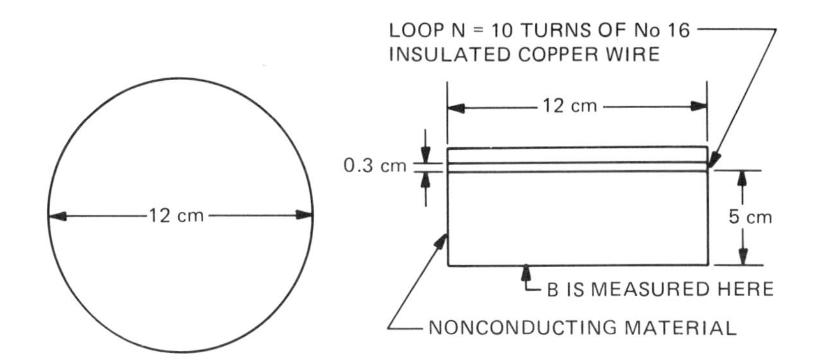

NOTES (1) $B = 5 \cdot 10^{-5}$ T/A at 5 cm from wire turns.
(2) Loop self-resonant frequency shall be greater than 100 kHz.

Fig 37
Loop Used for Radiating Magnetic Fields

4.2.7.3.2 Magnetic-Field Susceptibility Test [19]. The susceptibility to the magnetic field is generally measured in the frequency range of 30 Hz to 30 kHz. The following apparatus will be required for this test:

(1) Signal (noise) source
(2) EMI meter
(3) Radiating loop

As shown in Fig 37, the radiating loop is capable of producing a magnetic flux density of $5 \cdot 10^{-5}$ T/A at a point approximately 5 cm from the face of the loop. It is supported on a wooden form or similar nonconducting material.

Figure 38 shows the acceptable susceptibility limits for magnetic fields as required in [19].

4.3 Wiring Type Classification

4.3.1 Definition. For the purpose of this guide, wiring types are classified by their suppressive, barrier, and compensatory performance. These parameters describe only the effect on coupling (see 3.4) of a given wiring technique.

4.3.1.1 Suppressive Wiring Techniques. These techniques result in the reduction of electric or magnetic fields in the vicinity of the wires that carry current without altering the value of the current.

Wires which are candidates for suppressive techniques are generally connected to a noise source. They may couple noise into a susceptible circuit by induction, for example, twisting or transposing of ac power lines to reduce the intensity of the magnetic field produced by current in these lines.

4.3.1.2 Barrier Wiring Techniques. These techniques obstruct electric or magnetic fields, excluding or partially excluding the fields from a given cir-

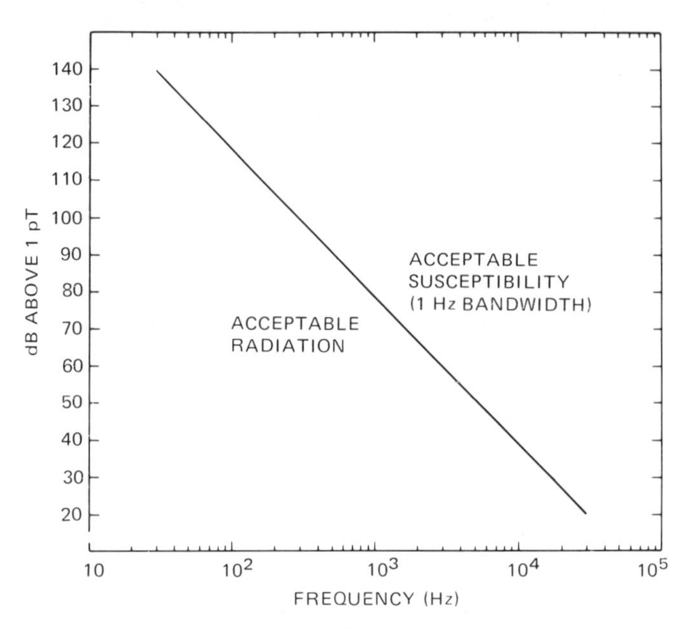

Fig 38
Limits for Radiated Emission and Susceptibility

cuit. Barrier techniques are often effective against electromagnetic radiation also (see 3.4.4). In general these techniques change the coupling coefficients between wires connected to a noise source and those connected to the signal circuit, for example, the placement of signal lines within steel conduit to isolate them from an existing magnetic field.

4.3.1.3 Compensatory Wiring Techniques. These techniques result in substantially canceling or counteracting the effects of the rates of change of electric or magnetic fields, without actually obstructing or altering the intensity of the fields. If the signal wires are considered to be part of the control circuit, these techniques change the susceptibility of the circuit, for example, the twisting of signal and return wires associated with a susceptible instrument so as to cancel the voltage difference between wires caused by an existing, varying magnetic field.

4.3.2 Classification. The suppressive, barrier, or compensatory behavior of a given wiring technique depends upon the field responsible for the noise (electric, magnetic), the frequency range, and often upon the nature of the load present on the affected wiring. Therefore the quantitative evaluation of a given technique must take all such parameters into account as they relate to a specific application. However, a qualitative classification of the various wiring techniques is possible, showing applicability and usual performance. Table 3 presents common examples.

Table 3
Qualitative Wiring Technique Classification

Wiring Technique	Usage Behavior Frequency	Electric-Field Decoupling						Magnetic-Field Decoupling					
		Suppressive		Barrier		Compen-satory		Suppressive		Barrier		Compen-satory	
		High	Low	High	Low	High	Low	High	Low	High	Low	High	Low
Common-ground signals		Poor	Poor	—	—	Poor	Poor	Poor	Poor	—	—	Poor	Poor
Return in signal cable		Poor	Poor	—	—	Poor	Poor	Fair	Fair	—	—	Fair	Fair
Paired conductors		Fair	Good	—	—	Fair	Good	Good	Good	—	—	Good	Good
Twisted pairs		Fair	Good	—	—	Fair	Good	Good	Good	—	—	Good	Good
Shielding, braided		Good	Good	Good	Good	Good	Good	Fair	Poor	Fair	Poor	Good	Good
Shielding, copper foil		Good	Good	Good	Good	Good	Good	Fair	Poor	Fair	Poor	Good	Good
Shielding, Mumetal		Good	Good	Good	Good	Good	Good	Good	Fair	Good	Good	—	—
Coaxial cable		Good	Good	—	—	Good	Good	Good	Good	—	—	Good	Good
Conduit, steel		Good	Good	Good	Good	—	—	Good	Good	Good	Good	—	—
Conduit, aluminum		Good	Good	Good	Good	—	—	Good	Poor	Good	Poor	—	—
Cable tray		Fair	Fair	Fair	Fair	—	—	Fair	Fair	Fair	Fair	—	—

4.3.3 Quantitative Evaluation. In an effort to illustrate the importance of suppressive, barrier, and compensatory wiring techniques, and at the same time to suggest an approach to an objective relative evaluation of various wiring approaches, a collection of test circuits is presented. Tests shown are concerned solely with electric and magnetic fields such as may be found in normal industrial environments, and which are likely to disturb low-energy control circuitry. Electromagnetic interference (radio interference) has not been considered.

All of the test setups shown (Figs 39–42) provide a reference standard configuration. The evaluation of a specimen wiring technique is performed by logging response voltage versus frequency for both the standard and the specimen, then calculating the ratio of the two at various frequencies. This ratio, taken at the poorest test frequency, then converted to decibels, represents the improvement due to the specimen wiring technique.

R and T should be selected to permit up to 100 V rms (open circuit) driver output and up to 1 A rms (short circuit).

The variable-frequency oscillator should provide a range of 60–1 000 000 Hz.

The oscilloscope should provide differential input at 100 dB common-mode rejection and a $10 \cdot 10^{-6}$ V/div deflection factor.

Fig 39
AC Voltage, Current Source and Readout

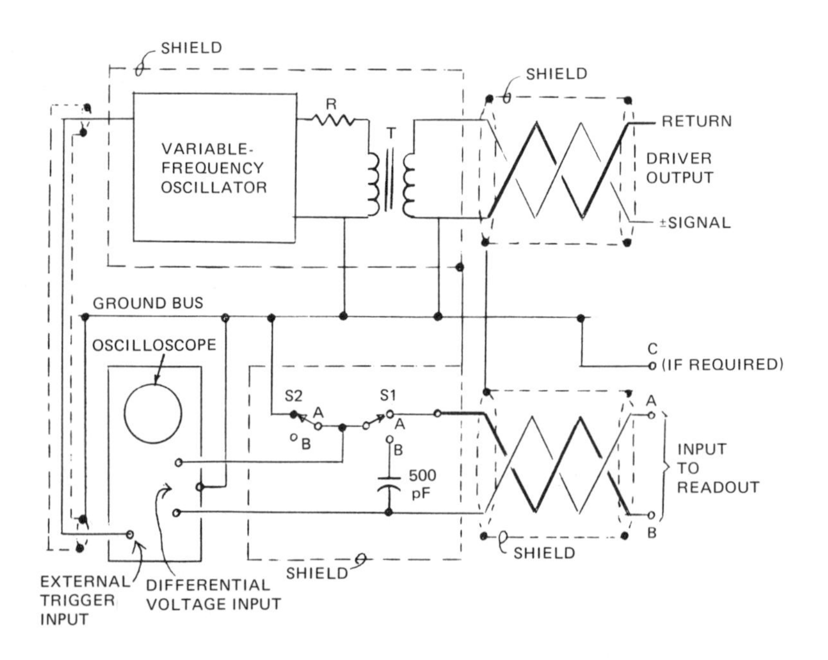

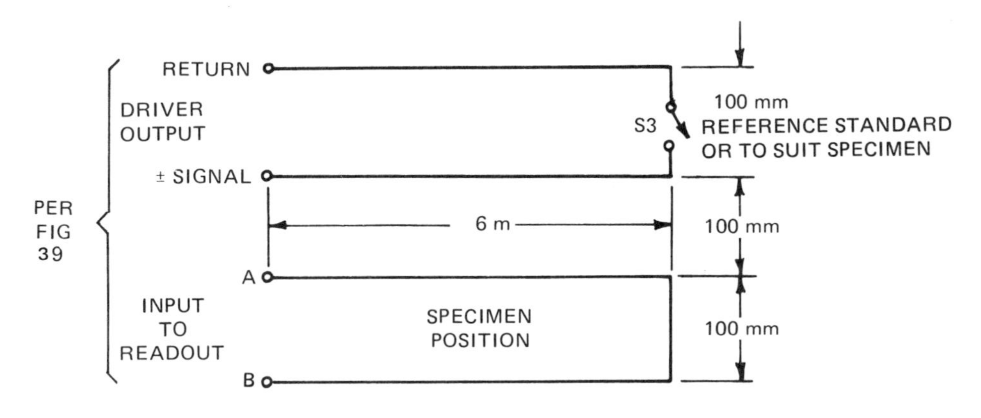

Fig 40
Suppression Evaluation Setup

(Set switches as follows: for magnetic suppression testing, S1 to a, S2 to a, S3 closed; for electric suppression testing S1 to b, S2 to a, S3 open.)

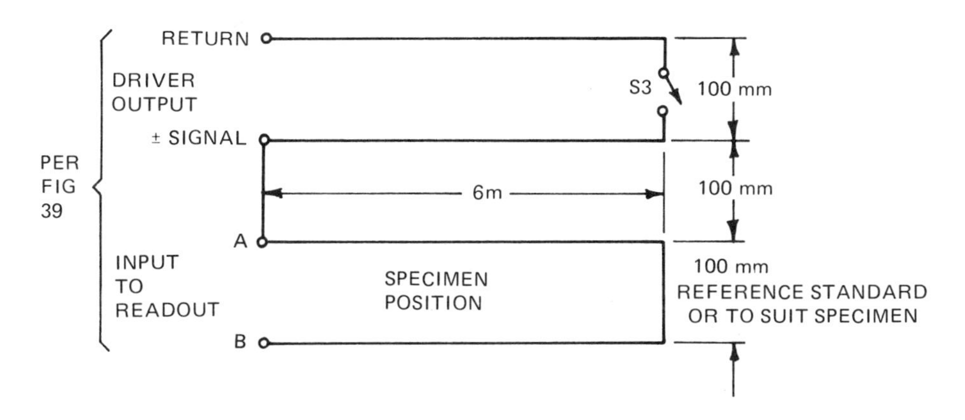

Fig 41
Barrier Evaluation Setup

(Set switches as follows: for magnetic barrier testing, S1 to a, S2 to a, S3 closed; for electric barrier testing, S1 to b, S2 to a, S3 open.)

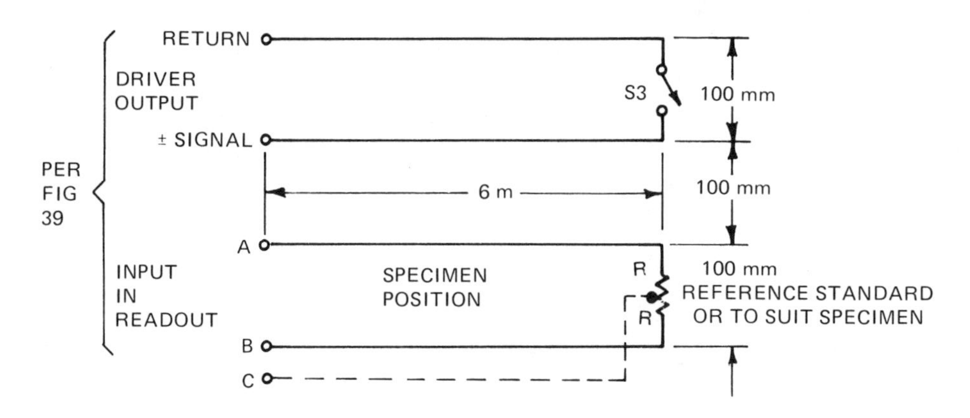

Fig 42
Compensation Evaluation Setup

(Set switches as follows: for magnetic compensation testing, S1 to a, S2 to b, S3 closed; for electric compensation testing, S1 to b, S2 to b, S3 open.)

The use of oscilloscope external triggering, as shown, is helpful for the identification of spurious noise pickup, but is not absolutely essential.

In Fig 42 specimen conductor A should be a signal line. Specimen conductor B should be the outer conductor for coaxial cable or a second signal line for paired cable. Specimen conductor C should be the shield for shielded pair cable or may be separate wire if no shield exists.

Resistor R typically should be 50 Ω, although for certain applications other values, including a substantial unbalance, are preferred.

4.3.4 Nonelectrical Signal Transmission. Other forms of signal transmission than the classical direct-wired method are possible. For instance, the use of optical "light pipes" and of acoustic signaling is occasionally observed. Such nonelectrical methods generally are unaffected by electrical noise and will not be considered further within this guide.

4.3.5 Installation Design Parameters

4.3.5.1 Wire Size. In selecting the wire size for making interconnections between different parts of a control system, the first consideration must be that the wire meet the current-carrying requirements for that application [10], Tables 310-12 to 310-15, 310-2(b)], [11], [16], [36], [39], [42], [47], [50]. The next considerations are the physical space available within the termination point and the physical strength necessary to prevent breakage which might occur during installation or due to an application where there is severe stress on interconnecting cables and wires.

Although these are the primary factors in selecting the wire size, there are others which should be considered to give the best noise immunity of interconnecting wires and cables.

Minimizing the conductor size will reduce the capacitance between adjacent insulated wires, reducing the electric-field coupling. Minimizing the wire size will also reduce the distance between conductors in a cable, resulting in a slight increase in magnetic-field cancellation.

Because the resistance per unit length increases as the wire size is reduced, consideration must be given to the length of the run so that the conductor resistance and impedance do not become an appreciable part of the circuit resistance or impedance.

4.3.5.2 Cable Insulation. The type of cable insulation is selected on the basis of the insulation voltage and the environmental requirements along with the physical space available and the physical strength requirement previously mentioned. Selecting a minimum diameter cable which will meet these requirements will result in the conductor centers being closer together, slightly reducing the magnetic-field coupling between conductors in a cable or raceway.

4.3.5.3 Wire Type. In 4.3.2 the various wiring techniques are compared as to their noise rejection ability. Section 6 recommends their proper usage. The intent of this section is to identify some of the types.

4.3.5.3.1 Twisted Pairs. A uniform twist of the signal and return wires provides one of the most effective means of reducing electrical noise due to magnetic-field coupling. Good noise cancellation is provided by 12–16 twists per foot. This is about twice as effective as twisted wires with a 3 inch lay. For critical circuits it may be desirable to further reduce the electric-field effects by adding a shield to the twisted pair. Copper or steel braid or copper or aluminum tape are frequently used as shielding. Cable shields should be insulated from each other and from ground and should be grounded at one end only.

4.3.5.3.2 Coaxial Cable. Coaxial cable consists of an inner and an outer conductor, insulated from each other, with both conductors carrying the desired signal currents. The outer conductor is usually grounded at the source. This is effective in protecting against magnetic and electric fields.

4.3.5.3.3 Triaxial Cable. Triaxial cable is coaxial cable with an additional outer copper braid insulated from the signal-carrying conductors acting as a shield. The inner conductor is normally used as the signal conductor, the inner shield as the signal return, and the outer shield as a true shield grounded at one end. Triaxial cable used this way improves the signal-to-noise ratio over that of standard coaxial cable.

4.3.5.3.4 Quadraxial Cable. Quadraxial cable is used for extremely low-level signals in noisy environments. The two inner conductors are twisted balanced wire having a specific impedance. There is an inner insulated braid around this twisted pair and an outer insulated braid. The inner braid is normally connected to equipment ground (grounded at one end), and the outer braid is connected to earth ground. If separate grounds do not exist, both shields are tied to earth ground.

4.4 Shielding. The objective of shielding is to protect a component, circuit, or system against the effects of undesirable external disturbing sources. The technique of shielding differs, depending upon the characteristics of the external disturbing source and the mode of transmission of its disturbances, such as low- or high-frequency disturbing sources and electric, magnetic, or electromagnetic fields. The transmitting field is usually characterized by the ratio of its electric-field component to its magnetic-field component. If this ratio is 377 Ω, the field is called an electromagnetic field; if the ratio is higher than 377 Ω, it is called an electric field, and if the ratio is lower than 377 Ω, it is called a magnetic field.

4.4.1 Shielding against Low-Frequency Fields

4.4.1.1 Electric Fields. For dc and low-frequency fields an electrostatic shield maintains the voltages of components inside it constant with respect to itself. If the interference source changes the voltage of the shield with respect to a reference (ground), then the voltages of the shielded components will also tend to change similarly with respect to the reference (ground). Therefore the voltage of the electrostatic shield should be held constant for effective shielding. In general, the electrostatic shield is held at ground potential. For higher effectiveness, the conductivity of the electrostatic shield should be as high as possible. Copper is widely used as the electrostatic shield. The principle of electrostatic shielding is illustrated in Fig 43.

Fig 43
Electrostatic Shielding

(1—shielded object; 2—shield; 3—disturbing external source.)

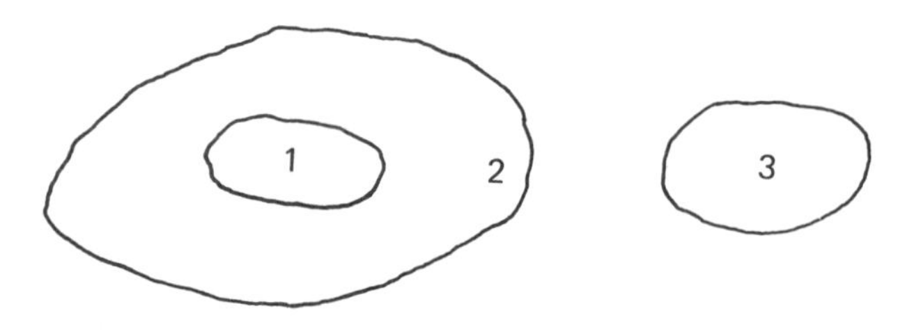

Assume that conductor 1 is completely surrounded by conductor 2, and that conductor 3 (disturbing source) is brought near them. If the charge of the *i*th conductor is q_i and the voltage is V_i, for the special case where $V_2 = 0$, we have

$$q_1 = c_{11} V_1 + c_{13} V_3$$

$$q_2 = c_{12} V_1 + c_{23} V_3$$

$$q_3 = c_{13} V_1 + c_{33} V_3$$

If $q_1 = 0$, then the potential inside the hollow is the same everywhere and is equal to that of the hollow itself, which is zero (because $V_2 = 0$), that is, $V_1 = 0$. Then

$$c_{13} V_3 = 0$$

or

$$c_{13} = 0$$

because V_3 can have any value. Therefore, for the more general case when q_1 is not zero,

$$q_1 = c_{11} V_1$$

$$q_3 = c_{33} V_3$$

In other words, the potential of each conductor except that of the shield is proportional to its own charge, and is unaffected by the charge of the other conductors. The potential of the shield must remain invariant for perfect shielding against external electric fields. In practical cases it is sometimes difficult to completely isolate a conductor from the others.

Any change in voltage on conductor 3 will induce a change in voltage on conductor 1 [Fig 44(a)]. Conductor 1 is said to be capacitively coupled to conductor 3. If conductor 2 is brought in the neighborhood of the other two conductors [Fig 44(b)], then conductor 2 will be capacitively coupled to both conductors 1 and 3. The voltage on conductor 3 (with respect to ground) will induce a voltage on conductor 2. Conductor 2, in turn, will induce a voltage on conductor 1. Therefore the effect of the isolated conductor 2 will be to indirectly increase the capacitive coupling between conductors 1 and 3. However, if conductor 2 is grounded (that is, connected to the reference point of conductor 3), much of the electric flux from conductor 3 will be shunted away from conductor 1 to conductor 2. In effect, this would decrease the capacitance (and the capacitive coupling) between conductors 1 and 3. In other words, the grounded conductor 2 would partially shield conductor 1 from conductor 3. Therefore it is good practice to tie all isolated and unused conductors to ground.

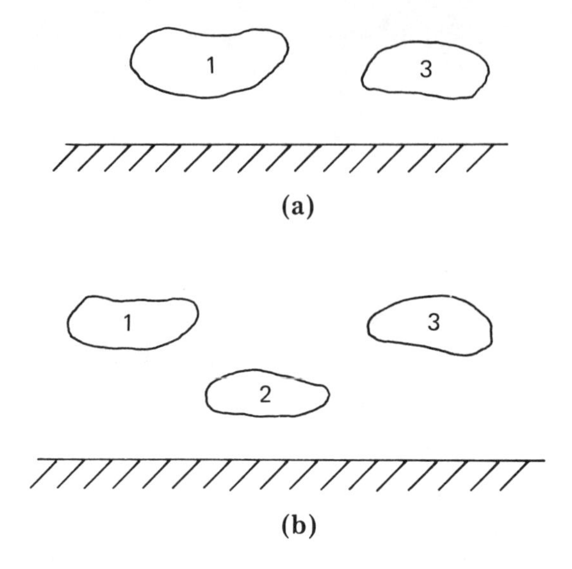

Fig 44
Partial Shielding

4.4.1.2 Magnetic Fields. For shielding against dc and low-frequency magnetic interference fields, the magnetic flux is diverted from the surrounding space to the shielding material. An effective magnetic shield should have low reluctance. Reluctance is inversely proportional to the permeability and the cross-sectional area of the shield. Therefore for higher effectiveness, the shield should have high permeability and a large cross-sectional area. Since magnetic material is nonlinear, the permeability depends upon the magnetic flux density inside the shield wall. Therefore the ambient magnetic flux density should be known in order to estimate the effectiveness of a given shield.

If the magnetic circuit consists of two parallel paths, one of permeability $\mu >> \mu_0$ and the other of μ_0, then most of the magnetic flux will be diverted to the path having permeability $\mu >> \mu_0$. This is the principle of shielding against dc and low-frequency magnetic fields. When a piece of apparatus is required to be shielded against stray magnetic flux, it is surrounded by a magnetic material of high permeability. Then most of the stray magnetic flux will be contained inside the shield. The magnetic flux in the wall of a spherical high-permeability shield can be estimated as [62].

$$B_s = \frac{2B_a}{t_{01}} \quad \text{T}$$

where

B_s = magnetic induction in shield, tesla
B_a = magnetic induction of ambient, tesla
t_{01} = shield thickness/outer radius

The attenuation of the magnetic field is most conveniently expressed in decibels (dB),

$$\text{attenuation} = 20 \log_{10} \left(\frac{B_1}{B_2}\right) \quad \text{dB}$$

where

B_1 = magnetic induction before shield is placed
B_2 = magnetic induction after shield is placed

For long cylindrical shields the attenuation is approximately given by

$$\text{attenuation} = 20 \log_{10} \left(\frac{\mu_r t}{d}\right) \quad \text{dB} \qquad \text{if} \quad \frac{\mu_r t}{d} >> 7.5$$

where

t = wall thickness of shield, m
d = outer diameter of shield, m
μ_r = permeability of shield material relative to that of free space

For a single thin spherical shell,

$$\text{attenuation} = 20 \log_{10} \left[1 + \frac{2}{3}\frac{t}{d} \cdot \frac{(\mu_r - 1)^2}{\mu_r}\right] \quad \text{dB}$$

For two concentric spherical shells separated by nonmagnetic spacers,

$$\text{attenuation} = 20 \log_{10} \left[1 + \frac{2}{3} \mu_r (t_{01} + t_{23}) + \frac{4}{3} \mu_r{}^2 t_{01} d_{12} t_{23}\right] \quad \text{dB}$$

where

$$t_{01} = \frac{t_1}{r_0}; \qquad t_{23} = \frac{t_3}{r_2}; \qquad d_{12} = \frac{d_2}{r_1}$$

and

r_0 = inner radius of innermost shell, m
r_1 = outer radius of innermost shell, m
r_2 = inner radius of outermost shell, m
r_3 = outer radius of outermost shell, m
t_1 = $r_3 - r_0$
t_3 = thickness of outermost shell, m
d_2 = thickness of nonmagnetic spacers, m

As mentioned earlier, the permeability of a magnetic material is not constant, but is dependent on the magnetic flux density (magnetic induction) in the shield material. If the ambient magnetic field is so high that the magnetic shield becomes saturated for the wall thickness chosen, then the shield will be practically ineffective. In any case, complete magnetic shielding is not possible. In contrast, in many cases effectively complete electrostatic shielding is possible and a previous knowledge of the ambient electric field is not required. Therefore electrostatic shielding is more effective and simpler than magnetic shielding.

4.4.2 Shielding against High-Frequency Fields [60], [69]. A high-frequency electromagnetic wave has, in general, both an electric-field and a magnetic-field component. Broadly speaking, there are three types of fields, plane-wave field, electric field, and magnetic field.

In a plane-wave field, the directions of the electric field and the magnetic field are perpendicular to each other and also perpendicular to the direction of the wave propagation. In addition, the magnitude of the electric field is the same at any point on a plane perpendicular to the direction of propagation. The same is true for the magnetic field. The ratio between the electric and the magnetic fields is called the wave impedance. For plane waves it is given by

$$Z_w = \sqrt{\frac{\mu}{\epsilon}}\ \Omega$$

where μ and ϵ are the permeability of the permittivity, respectively, of the medium. In free space the wave impedance of a plane-wave field is approximately 377 Ω.

The electric field is characterized by its large electric-field component compared with its magnetic-field component. The electric field is also called the high-impedance wave because its wave impedance is high. At large distances r from the source, when $r >> \lambda$ (λ being wavelength), the characteristics of the electric field approach those of the plane-wave field.

The magnetic field is characterized by its large magnetic-field component compared with its electric-field component. It is also called the low-impedance wave. For $r >> \lambda$, like the electric wave, the characteristics of the magnetic wave approach those of the plane wave.

When an electromagnetic wave impinges on a shield, a part of the wave is reflected by the shield wall. The amount of reflection depends upon the mismatch between the intrinsic impedance of the incident wave and the intrinsic impedance of the shield material. The rest of the wave is transmitted along the shield thickness. The transmitted part of the electromagnetic field is attenuated (absorbed) as it travels along the shielding material. As a result, the interference field, as it emerges from the other side of the shield, is lower in intensity than the original incident field.

4.4.2.1 Electric Fields. For static and low-frequency electric fields, a conducting surface enclosing the susceptible equipment provides perfect shielding, provided this shield is held at a constant potential (for example, ground).

There is no perfect shield against high-frequency electric fields. Upon impinging on the shield surface, some part of the incoming wave is reflected back and the rest travels through the shield, being absorbed (attenuated) by the shield as it travels along. The shielding effectiveness S is expressed as

$$S = A + R \quad \text{dB}$$

where

A = absorption loss
R = reflection loss

NOTE: The quantity B, the rereflection loss, in [69] is being omitted in this presentation because it is usually a small correction.

The absorption loss can be calculated as follows:

$$A = 0.1314\ z\sqrt{FG\mu_r} \quad \text{dB}$$

where

z = thickness of shield, mm
f = frequency of electric wave, Hz
G = $g_s/5.8 \cdot 10^7$
g_s = conductivity of shield, S/m
μ_r = $\mu_s/4\pi \cdot 10^{-7}$
μ_s = permeability of shield, H/m

The reflection loss can be calculated as follows:

$$R = \begin{cases} 168 + 10 \log_{10}\left(\dfrac{G}{f\mu_r}\right) \quad \text{dB}, & \text{for } r >> \lambda \\ 354 + 10 \log_{10}\left(\dfrac{G}{f^3 \mu_s r^2}\right) \quad \text{dB}, & \text{for } << \lambda \end{cases}$$

where r is the distance of the disturbing source, in meter.

A shield of high conductivity is very effective for shielding against electric fields.

4.4.2.2 Magnetic Fields. For high-frequency magnetic fields, shielding is provided by reflection loss at the shield surface and absorption loss within the shield material. The absorption loss can be calculated from 4.4.2.1. The reflection loss can be calculated as follows:

$$R = \begin{cases} 168 + 10 \log_{10} \left(\dfrac{G}{f\mu_r} \right) \text{ dB}, & \text{for } r >> \lambda \\ 20 \log_{10} \left[0.138r \sqrt{\dfrac{Gf}{\mu_s}} + 0.354 + \dfrac{0.46}{r} \sqrt{\dfrac{\mu_s}{Gf}} \right] \text{ dB}, & \text{for } r << \lambda \end{cases}$$

4.4.3 Solution by Universal Curves [56]. In many cases it is convenient to calculate the shielding efficiency with the help of universal curves. Figure 45 shows the absorption losses (in neper) in a spherical metallic (magnetic or nonmagnetic) shell, where r_0 is the inner radius, d is the thickness, μ_r is its relative permeability, and δ is the skin depth. Figures 46 and 47 show the absorption-loss curves for magnetic and nonmagnetic cylindrical shields. Figure 48 is the nomogram for calculating the skin depth δ. Figures 49 and 50 are the universal curves for shielding against pulses when the shields are nonmagnetic and magnetic, respectively. The nomogram shown in Fig 49 is based on [73, eq (28)]. The following assumptions are made (Fig 49):

(1) The coaxial tubes are made of solid copper, without holes or slots in the outer tube.

(2) The radius of the outer tube is much larger than its thickness, $r_0 >> d$.

(3) The rectangular pulse on the outer tube has amplitude I_p and width T; the maximum current on the inner conductor is I_c.

If the material is not copper, multiply d by $\sqrt{p_{Cu}/\rho_m}$, which is 1.28 for aluminum, 2 for brass, 3.6 for lead, and 6.5 for stainless steel.

As an example, calculate the pulse width T that cannot be exceeded if the induced inner current I_c should be much smaller, say, −10 N, than the outer current I_p, when the thickness of the copper shield d = 0.1 mm (Fig 49). The answer is 20 ns.

In Fig 50 the limit theory of impulse shielding is based on the switching mechanisms (domain wall jumps, Barkhausen discontinuities) of ferromagnetics. For a mathematical derivation see [73].

The double nomogram shown in Fig 50 is designed for 80% Ni–Fe alloy (Supermalloy) having the following properties:

$$B_s = 0.75 \text{ Wb/m}^2$$
$$\rho = 1.6 \cdot 10^6 \ (\Omega \cdot \text{m})^{-1}$$
$$H_c = \begin{cases} 0.01 \text{ Oe (annealed)} \\ 0.4 \text{ Oe (work-hardened)} \end{cases}$$

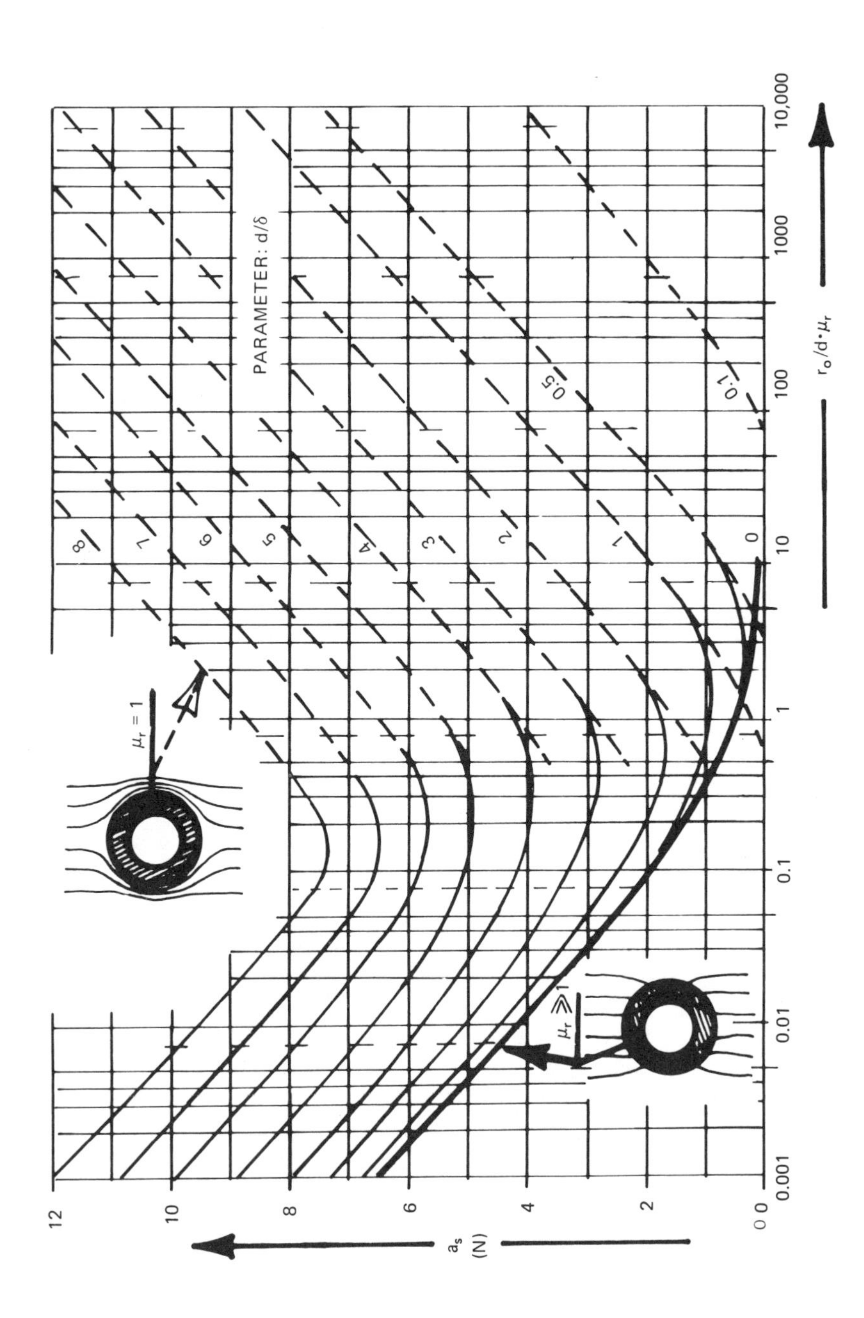

Fig 45
Shielding by Magnetic (Left) and Nonmagnetic (Right) Spheres

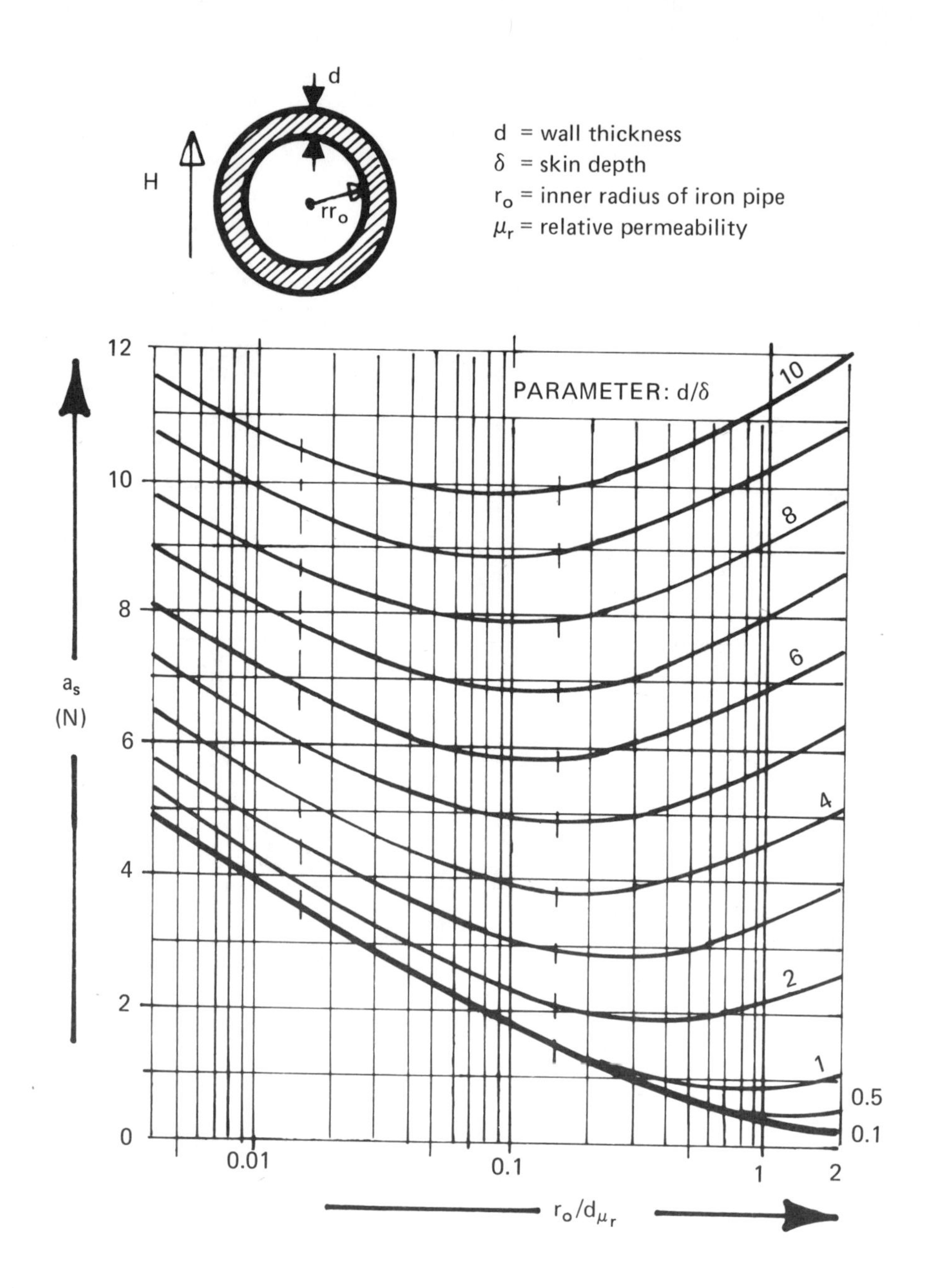

For cylinders replace x_0 by $r_0/2$
For spheres replace x_0 by $r_0/3$ (nonmagnetic only)

Fig 46
Shielding by Magnetic Cylinders

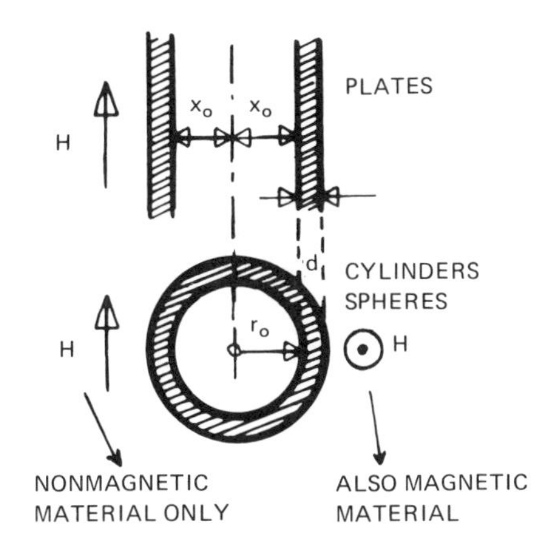

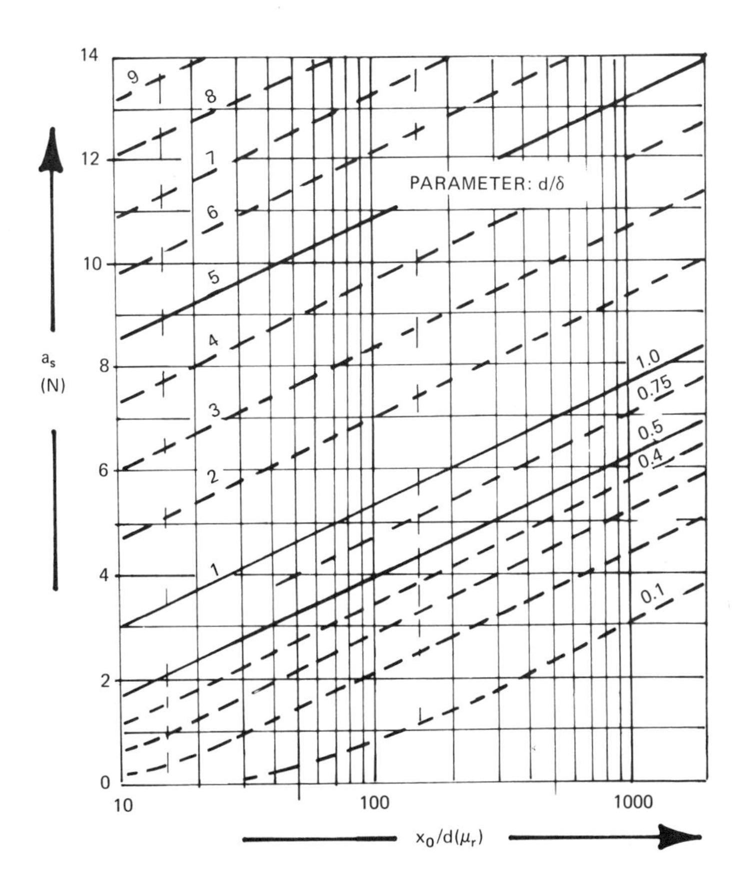

Fig 47
Shielding by Nonmagnetic Structures

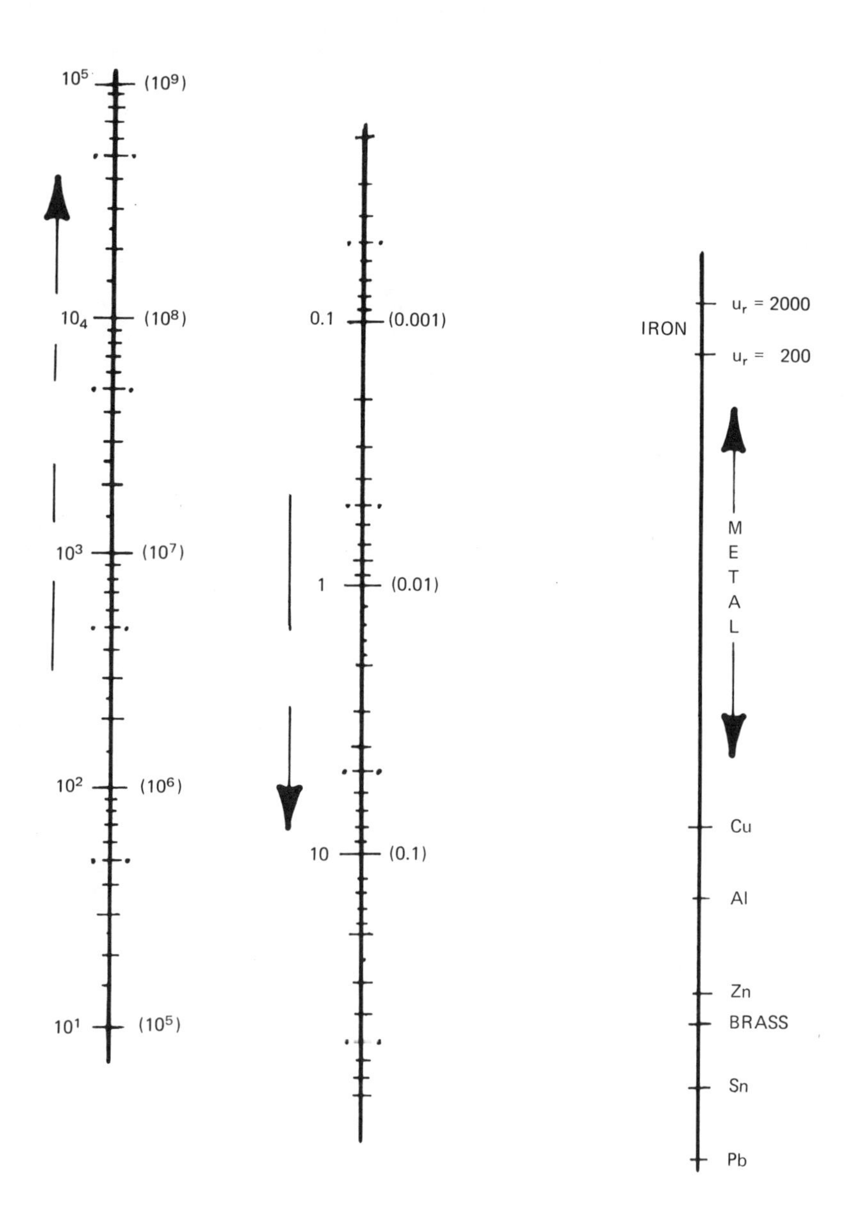

Fig 48
Equivalent Skin Depth

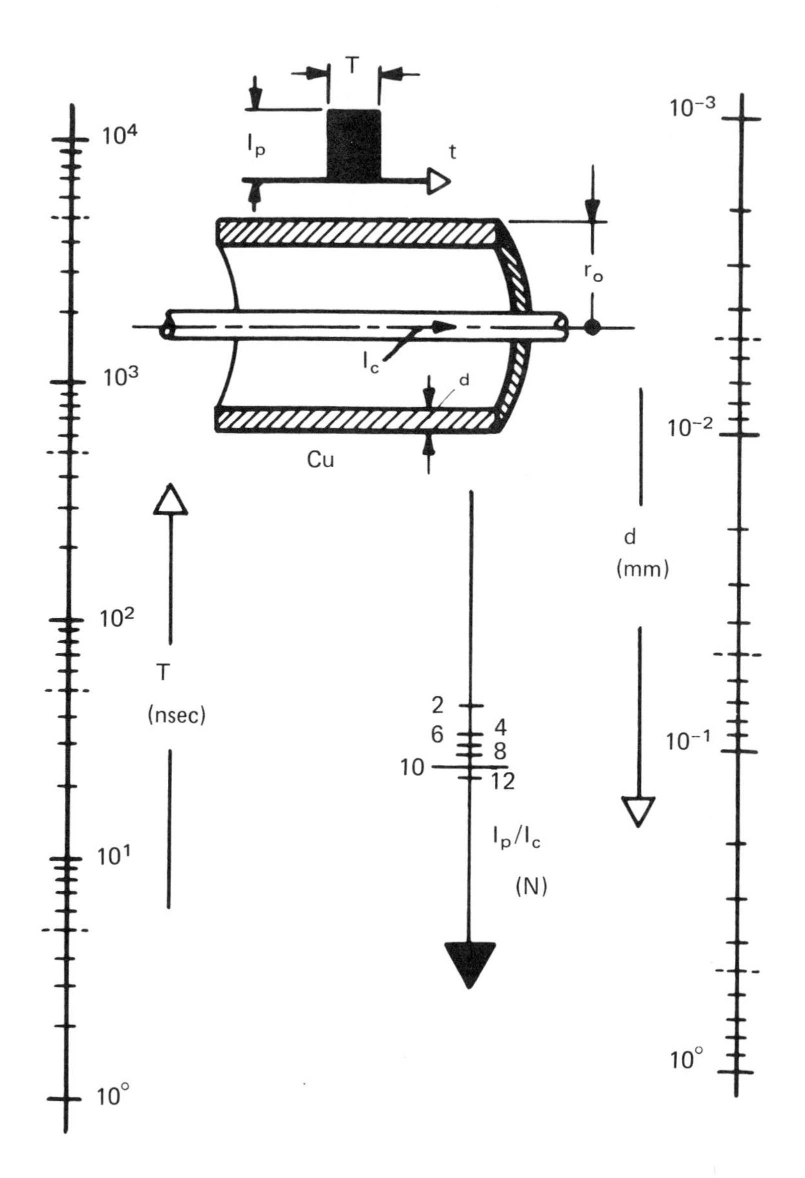

Fig 49
Impulse Shielding, Nonmagnetic Case

Note the steep transition between excellent shielding and negligible shielding.

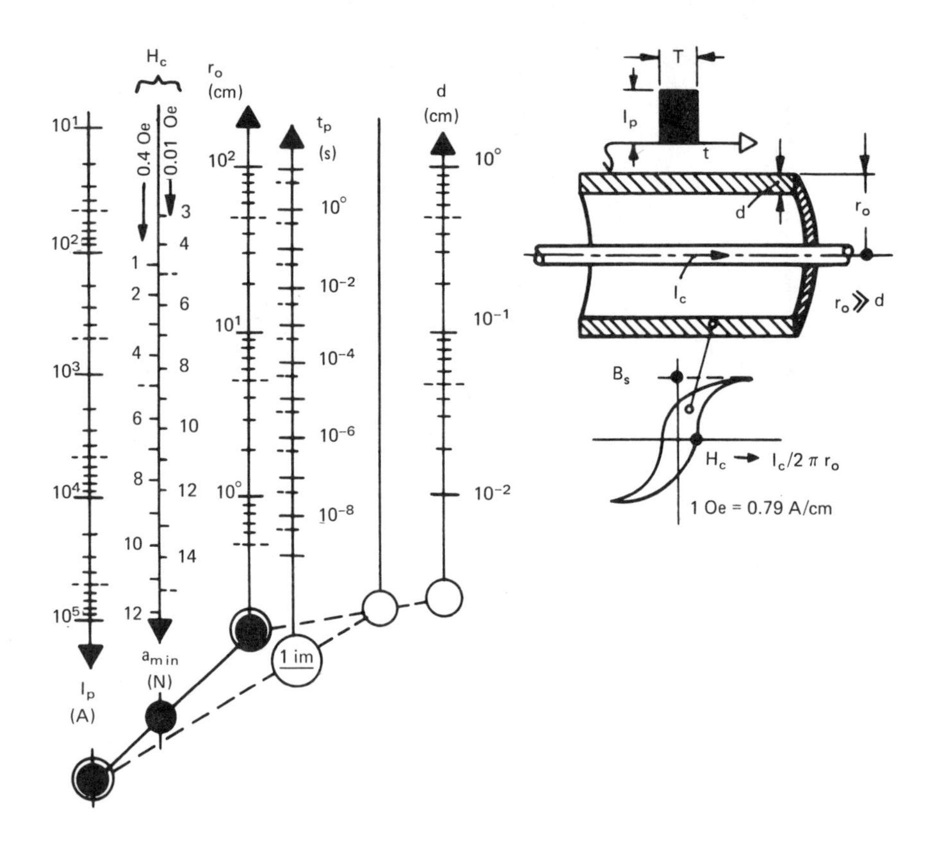

Fig 50
Impulse Shielding, Magnetic Case

For other than Supermalloy material use multipliers according to the design formulas.

$$\frac{I_p}{2\pi r_0 H_c} = e^{a_{min}}$$

$$t_p = \frac{\pi r_0 d^2 \rho B_s}{I_p}$$

For instance, for 41% Ni-Fe alloy, having a B_s = 1.54 Wb/m^2, multiply the protective time t_p by 2.

The minimum attenuation a_{min} occurs if d is such that the shell is just saturated throughout, which occurs when the pulse width T is equal to the protective time t_p. If $T >> t_p$, the attenuation is determined by only one material property, the conductivity.

4.5 Filtering and Buffering

4.5.1 Introduction. Filtering is used to eliminate conducted interference. Since the end effect is the same, namely, getting rid of noise, we shall not distinguish between filtering and buffering. The difference is only that a buffer uses media other than electric circuits which characterize the electromagnetic filter.

Filtering (and buffering or isolating) for the elimination of electrical noise must be considered under intrinsically broader aspects than conventional frequency-selective filters designed for information handling. Two generic aspects are involved.

4.5.1.1 Technical Aspects

(1) Interference filters are often subjected to much higher power than conventional filters. Since, for instance, power-line filters have to carry through the power, they often will be quite a bit larger, and the power may bias the (possibly nonlinear) filter elements (saturation).

(2) Often the power spectrum of the noise overlaps with the power spectrum of the power, control, or signal.

(3) The design of communication filters, as all filter books assure, is premised on impedance matching. In power feed lines in particular, this is not possible, since power feed lines are solely designed to be efficient at the power frequency and for nothing else. Thus mismatch often plays a very detrimental role, namely, a drastic reduction of the claimed or expected filtering and, quite often, the occurrence of pronounced ringing. Conventional filter design methods are, therefore, of very limited utility for noise elimination, and in the absence of impedance matching, Butterworth, Bessel, and other types of filters become rather inappropriate.

(4) High-peaked impulse noise combines the high energy of the noise with a very broad frequency spectrum.

4.5.1.2 Economical Aspects. There are many alternative avenues to clear interference. The decision of the most reasonable means must be based on the maximum benefit-to-cost ratio, with the benefit possibly being not much more than what is necessary. No specific single rule can be given for the decision involved, since it depends on the circumstances of the particular system under consideration and its noise environment. Rather, the reader should be familiar with all aspects of interference elimination and suppression. Filtering and buffering are quite often the most economical remedy. Filters are the only means to eliminate interference once it is on the line, but also the introduction of a filter or buffer close to the source can save on costly separation or wiring or on shielding. Buffers in particular, such as electrooptical isolators, are quasi-filters, which prevent the conduction of certain modes of noise, ground-loop induced or otherwise.

The classification of filters and buffers, in this practical context, will not be made on the basis of operating principles, but rather on the basis of those key properties that characterize the relationship of the noise on the one hand and the power, control, or signal to be conducted on the other hand. Typical, therefore, are the differences in power spectra or in amplitudes. This section, then, treats what is missing in conventional filter design methods.

4.5.2 Frequency-Domain Filters

4.5.2.1 Real Filter Elements

4.5.2.1.1 Real Capacitors (used by themselves for "high"-impedance loads). In many instances capacitors, in conjunction with inserted or already existing inductors, are shunted across the line for filtering purposes. This simple measure does not always work too well. Figure 51 shows reasons for and Fig 52 shows the effects of the nonideality of the real capacitors, which are often mistakenly assumed to be rather ideal two-terminal elements. Six differences are often overlooked:

(1) If the capacitor is not built as a feed-through capacitor, capacitative and inductive coupling between input and output leads, particularly at higher frequencies, may make the noise too circumvent, and hence nullify the effect of the capacitor. This feed-through requirement applies equally to filters which must also be mounted through the shield to prevent capacitive coupling at higher frequencies. If no shielding is available, the input (and possibly the output) line must be shielded.

(2) Another detrimental effect occurs when a not-feed-through arrangement is applied. Although mostly very small (but in terms of the small $1/\omega C$, it is not small enough), the series L in the shunt branch causes series resonance of the capacitor, above which the capacitor behaves like an inductor (Fig 52, curves A_1 and A_2).

(3) Internal shielding of the layers of wound capacitors reduces the ideal capacitance. Curve B of Fig 52 deviates from the ideal capacitor behavior characterized by the straight line of 20 dB per decade of frequency. This behavior is typical of paper, Mylar, and ceramic multilayer capacitors.

(4) The series resistance, at the small reactances occurring at high frequencies, can become dominant. Curve C of Fig 52 typically represents tantalytic capacitors which deviate drastically from the expected capacitive behavior. As curve C indicates, this can happen already at quite low frequencies. Hence tantalytic capacitors, although excellent at low frequencies, must be complemented by high-frequency capacitors to be shunted in parallel.

(5) At very high frequencies, ceramic capacitors (very-high-dielectric-constant capacitors) are preferred because of size, but internal resonances caused by transmission-line-effects may reduce their effectiveness. Discoidal (dish-like) feed-through capacitors (curve E) are quite a bit better than tubular ceramic capacitors (curve D). By splitting ceramic capacitors into two portions and providing a ferrite bead on the interconnecting wire, the capacitors can be made better (curves F_1, F_2, and F_3) than even ideal capacitors. (The

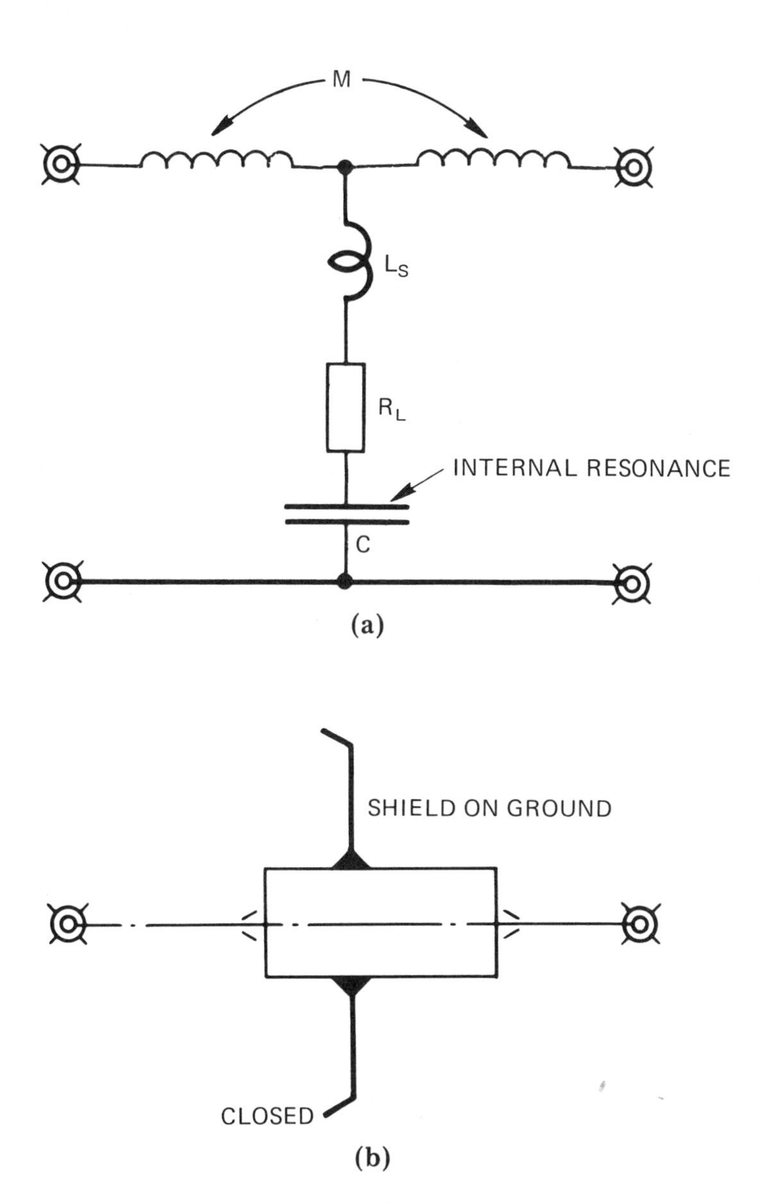

Fig 51
Reasons for Nonideality of Real Capacitors (a) Equivalent Circuit of Shunting Capacitor, Not Constructed as Feed Through. (b) Feed-Through Configuration

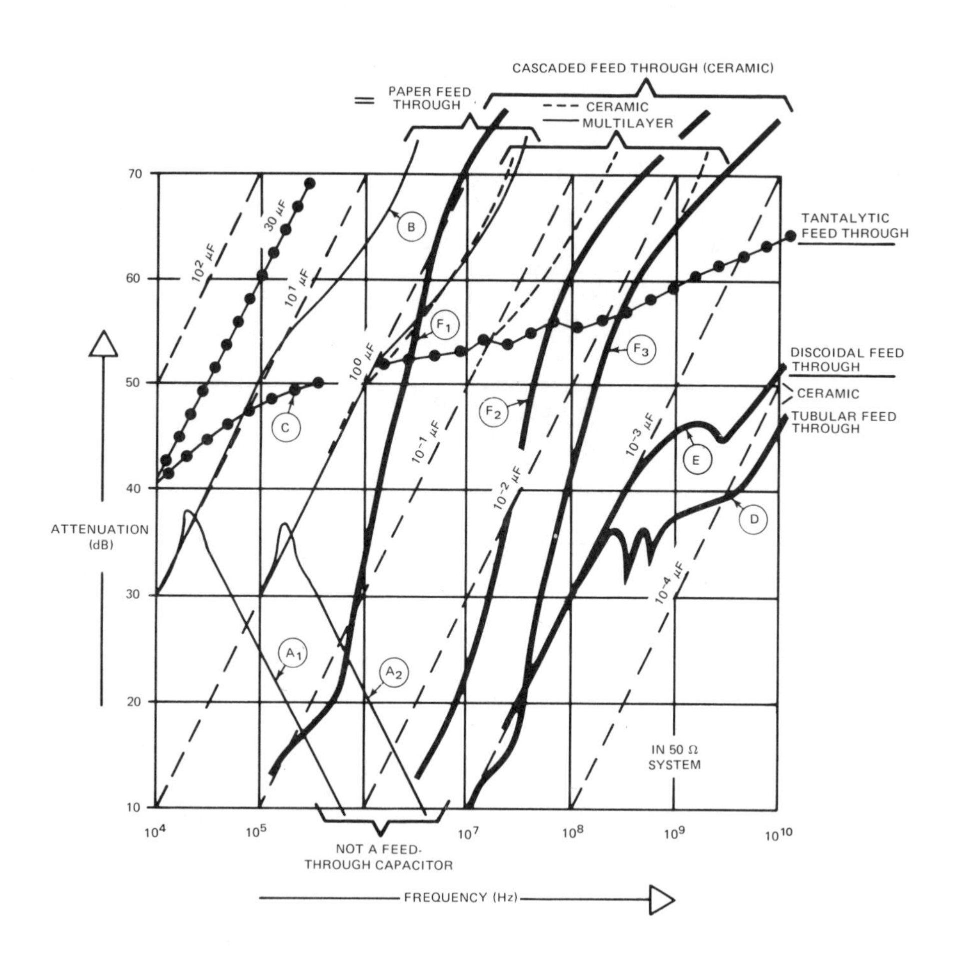

Fig 52
Attenuation of So-Called Capacitors

slope in Fig 52 is larger than 20 dB per decade of frequency.) Again, this is premised on feed-through configurations [58].

(6) Ceramic capacitors, when using very high dielectric constants and if improperly designed, may show voltage bias effects (a reduction of the effective dielectric constant by up to 70–80%) similar to the current-induced bias effect of inductors having magnetic cores (iron or ferrite).

4.5.2.1.2 Real Inductors (used by themselves for "low"-impedance loads). Ideal inductors do not exist just as ideal capacitors do not exist. An inductor, if iron or ferrite cored, may lose quite a bit of its inductance by saturating currents. This is quite pronounced in power feed line filters. It must be compensated by air gaps. (An exception are common-mode filters. See Fig 59.)

Real wound inductors with increasing frequency alternate between inductive and capacitive effects. Lossy inductors can be made of thick iron laminations at the expense of lowered inductance.

There exists a highly useful cheap inductor which can be used from about 1 MHz on, namely, the ferrite bead or block. One such bead, strung on a wire (one turn), represents an impedance of 20-50 Ω, with about 45° phase angle, for all frequencies above 1 MHz, and it does not "exist" (is negligible) at low frequencies. Current bias effects can be minimized by proper ferrite selection or by cracking the ferrite beads and gluing them together again (air gap).

4.5.2.2 Power Feed Line Filters. Here we must distinguish between three basically different cases:

(1) The interface impedances are known over the whole frequency range. Then we select and place filters such that a low impedance faces an L and a high impedance faces a C.

(2) The interface impedances are not known, but are time invariant. Then a multisection filter is the right answer.

(3) Sources or loads are being switched, either randomly or periodically. This is the dominant case in many an installation. But it is also the most misunderstood, frustrating, and confusing situation [56].

LC filters are presumed to be quite familiar to electrical engineers and may seem hardly worth reconsideration. Yet the very confidence in well-established filter theory is badly shaken if one finds that filters often do not work as predicted.

There is nothing wrong with established filter theory. [In this context it is tacitly assumed that the biasing (saturating) effect of power voltage and current has been reduced by proper design.] However, one must recall that filter theory is premised upon impedance matching. Yet, impedance matching does not at all exist in signal and control lines, and much less so in power feed lines. Power feed lines have the sole purpose of transferring power with high efficiency from the power source to the load. For lack of a better method, filter performance is at present being measured according to [17], that is, it is measured in a 50 Ω system. The actual system, however, in which the filter is supposed to work consists of source and load impedances, interface impedances which seem to vary "all over the map" as a function of frequency, location, and time. Figure 53 portrays the spread of actual filter performance as caused by variations of the interface impedances encountered. It may be +30 dB to −40 dB, in some cases even quite a bit more, and different from the value predicted in [17].

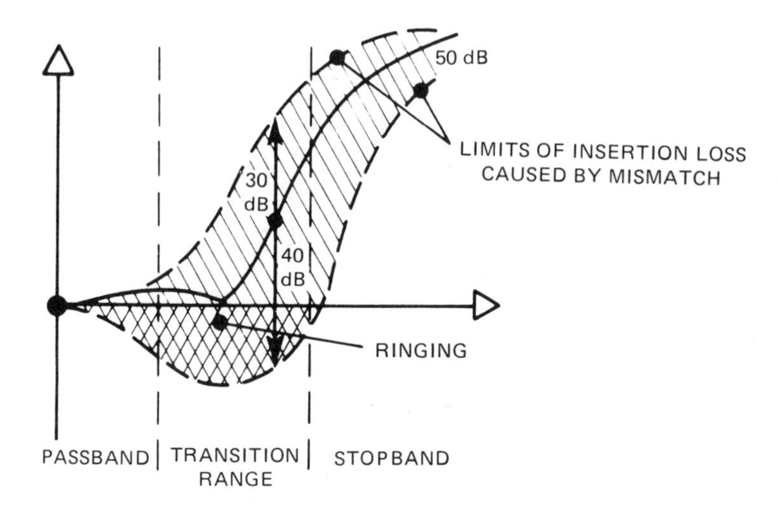

Fig 53
Effects of Mismatch

The shaded area in Fig 53 is bounded by the upper and lower limits of insertion loss (attenuation under the influence of mismatch). At high attenuation levels the bounds of insertion loss do not differ much from the attenuation curve. Hence for high attenuation, the filter prediction made in a 50 Ω system can be considered somehow realistic. But it is the transitional range and the passband where mismatch can play a deleterious role. Even insertion gain, or a negative insertion loss, can result because of interfacial and eigen(self) resonances.

However, the switching of loads or sources often has a much worse effect, namely, ringing of the filter. Such resonances can be quite annoying. For instance ringing, normally in the audio or ultrasonic frequency range, can be so pronounced (Fig 54) that SCRs misfire. It is often advisable to include small resistors (preferably carbon composition resistors for long-term stability under pulse conditions) to dampen such resonances. Lossy inductors are suitable, too. Unfortunately lossy filters for lower frequencies are not yet commercially available.

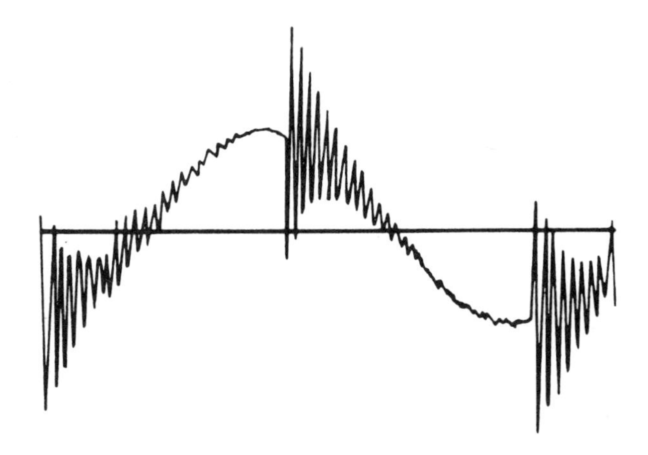

Fig 54
Typical Ringing

Much has been written on how badly conventional filters can behave under nonmatched conditions, but not much has been said on what to do about it in an economical and reliable way. Some people put so-called line-stabilization networks in the system, but for measurements purposes only. The actual filter operation is without line-stabilization networks, making the whole affair quite illusory. Others use brute force filters, heavily overdesigning filters, which results in rather costly and bulky filters. Again, others propose in-situ trial of filters [33], [34], which means that one tries for each filtering situation all kinds of filters until one finds, hopefully, one that works best—a tedious and nonoptimized affair without any chance of predictive planning. Others propose worst case filters [58], [59] having a guaranteed minimum insertion loss which operate for all possible interface impedances—a costly affair resulting in large filters.

It is stipulated in [18] that a 10 μF capacitor be inserted for measurement purposes (short circuit), but not left in for actual operation. Again this results in a great difference of assumed and actual performance of the filter.

The seemingly complex problem of filter dysfunction (righting in the passband and questionable performance in the stopband) has recently been clearly resolved [56, Chps 8 and 12].

4.5.2.2.1 Lumped-Element LC Filters. These are basically reflective filters, which reflect, but do not absorb, the unwanted interference. Such filters are indispensable if we must filter well below 1 MHz and, as is the case in complex systems, if we encounter variant and indeterminate interface impedances.

In addition to unavoidable and drastically changing mismatch, such filters (1) may ring or show strong insertion gain, or (2) may not filter as claimed, or both. Prevention of (2) is handled by multisection filters, and by introducing losses in the transistor band between passband and stopband. An efficient way to do this is to introduce eddy current losses selectively, such that the series inductors have Q's only where needed, normally in the range of 10—100 kHz. Figure 55 depicts how this is done. A winding goes around two cores of the configuration, which are of different heights. The higher core L_1 has heavy laminations of a thickness d twice the skin depth at corner frequency f_1, above which we need a low Q. The upper end of the low-Q region is given by the frequency for which L_2, imparted by the fine-lamination core and increasing proportional to f, overtakes the L_1, of the heavy core, which increases only with the square root of f. In this context we must mention that the upcoming CISPR and ANSI standards on mismatched filters use the term "worst case." Worst case, for many an engineer, implies that it happens only seldom and hence is not very critical. The truth is that most complex modern systems, in particular control systems, by their very nature have continually changing interface impedances. For such virtually normal conditions, filters meeting uncritical military standards [17] or other outdated standards can do more harm than good. For more details see [56].

4.5.2.2.2 Commercial-So-Called Lossy Filters or Lines. The filters, which are not to be confused with the band-limited loss filters just described, are very good for eliminating interference above 10 MHz. However, since they have no significant losses below about 10 MHz, they also result in insertion gain, ringing, or both, mostly in the megahertz range.

4.5.2.3 RC Filters: Passive, Active, and Quasi

4.5.2.3.1 Passive RC Filters. In control systems 60 Hz energy permeates the whole environment. If it has to be removed from sensitive signal lines, a stable twin-TRC band-reject network may often be sufficient to notch out the disturbing 60 Hz.

4.5.2.3.2 Active Filters. For signal lines active filters combining resistances, capacitances, and integrated operational amplifiers replace effectively LC filters which would be too large at low and very low frequencies, even at signal line levels, without appreciable bias. A well-established literature exists on active filters, the most comprehensive presentation being [45].

Although [58] describes an active 60 Hz power feed line filter, and although such filters have been built, it seems more economical and more adequate to apply filter regulators, which will be described in 4.5.3.1. Active dc line filters are essentially power regulators. Typical examples and conditions to be observed are given in [58]. From this source Fig 56 is reproduced, which shows the basic diagram of a dc line filter which reduces randomly switched load current pulses on the order of 10 ms and several amperes to wiggles in the milliampere range on the supply side.

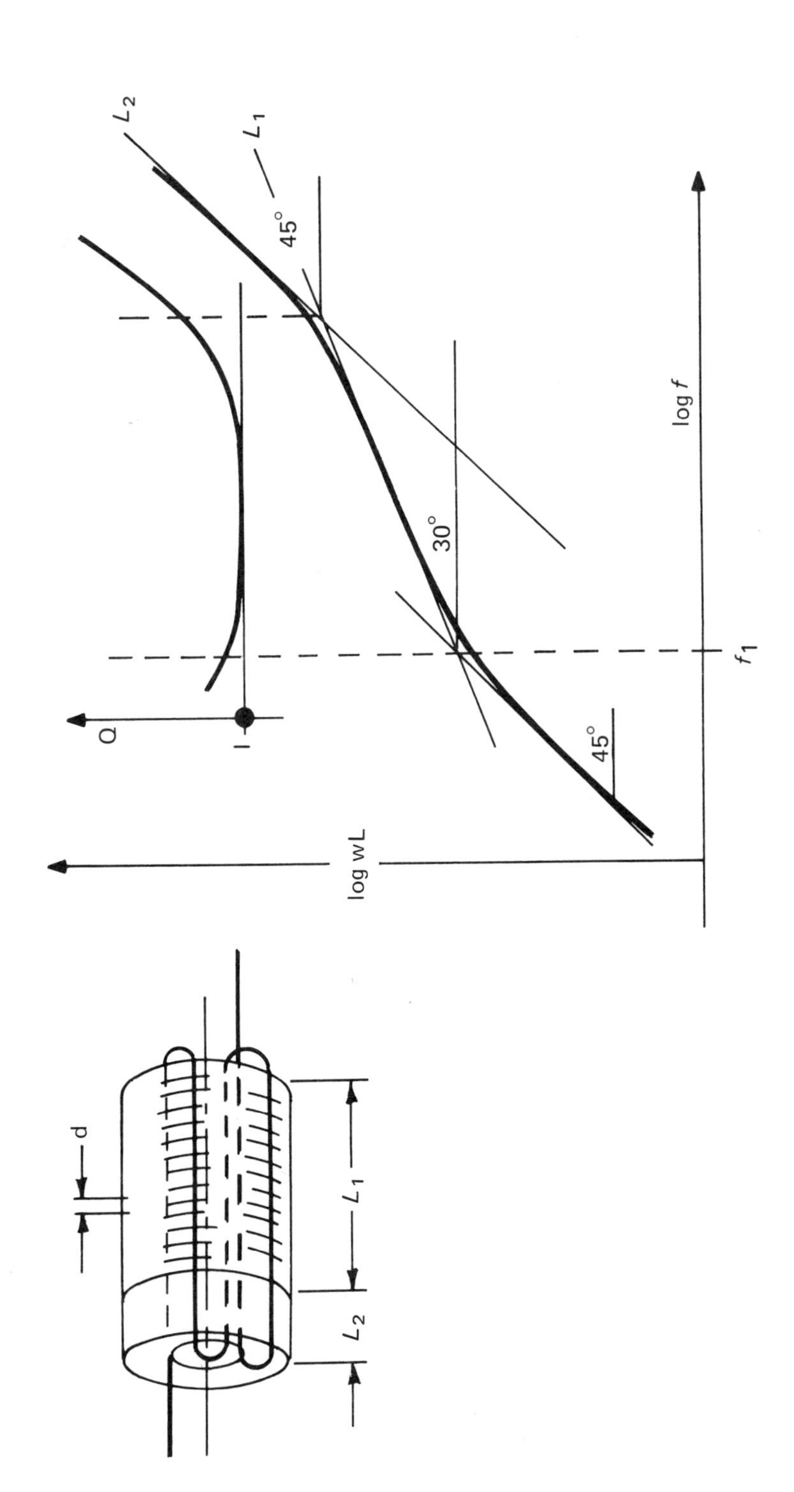

Fig 55
Inductor Having Band-Limited Losses

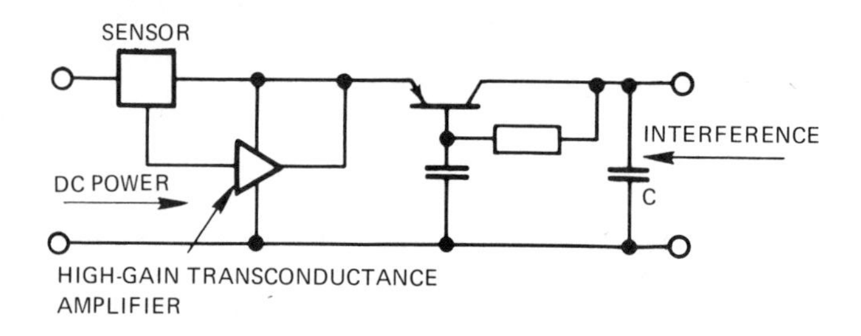

(a)

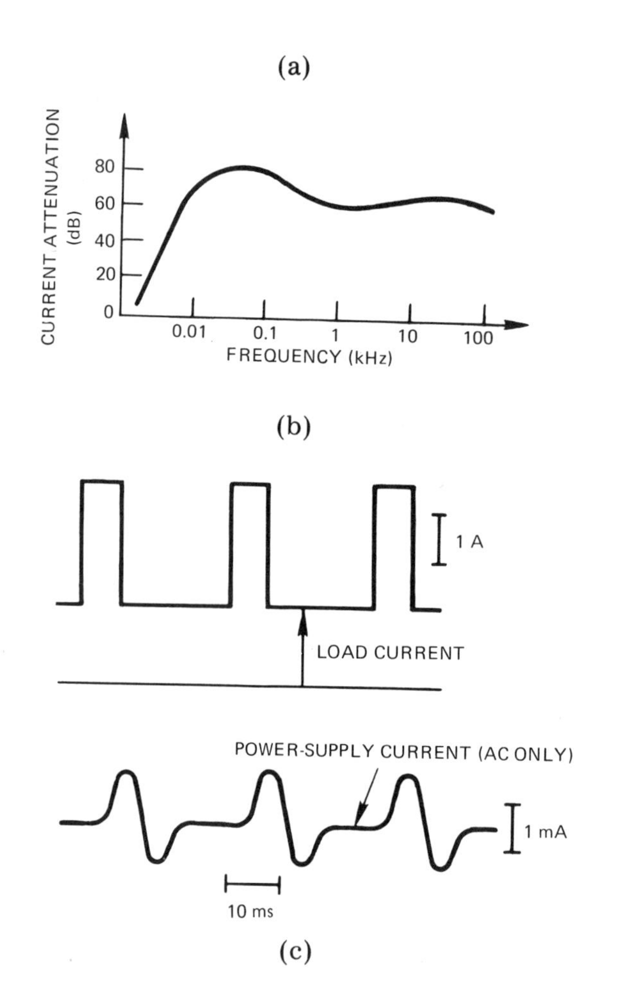

(b)

(c)

Fig 56
Random Pulse Filter. (a) Basic Circuit for Wideband Attenuation. (b) Current-Attenuation Curve. (c) Attenuation of 20 Hz, 2.5 A Current Pulses

4.5.2.3.3 Quasi-RC-Filter Switches. Electrooptical isolators [Fig 60(a)] for which the light path is interrupted or opened by a movable, isolated obstruction and for which the light sensor consists of slowly reacting cadmium sulfide, act as bounce- and interference-free switches having moreover the advantage of isolation.

4.5.3 Other Filtering Approaches

4.5.3.1 Filter Regulators. Switching loads may cause severely disturbing transients. The power spectrum of transients contains such low frequencies of high amplitude that conventional LC filters would have to be of unmanageable size to be working efficiently. In dc lines active filters, as previously outlined, can reduce this interference to quite acceptable levels. Quite often simple energy storing elements are adequate as, for instance, tantalytic capacitors or secondary batteries. But even in logic circuits [since switching times are on the order of nanoseconds and can, even for small L, cause large $L(di/dt)$] energy storage in the form of simple ceramic capacitors, like chip capacitors, across the dc line, close to the gates, can eliminate undesirable transients in shared power supplies. In ac lines storage elements are tuned circuits, which are rather unwieldly arrangements at 60 Hz. Since storage elements are much easier to provide for, any direct current, rectification, storage, and conversion back to alternating current (for instance, by oscillator amplifiers, switching, or inverter types) can stiffen power supplies quite drastically. Such devices also render good filtering, but are usually quite expensive. Hence for low cost, often ferroresonant transformers are inserted in the line. They are not only modest in price, but also modest in performance, unless properly modified. Figure 57, taken from [57], marshals the regulation and filtering performance of three types of ferroresonant transformers. α is an optimized ferroresonant filter regulator having excellent regulation and filtering, in contrast to the conventional ferroresonant transformer γ. β is another ferroresonant transformer, having high reactive current and poorer regulation and, also, as does β, behaving very poorly (collapsing) by half-cycle interruptions. [Curves with half-cycle interruption, as shown in Fig 57(e), mean that half the cycle is clipped off the primary voltage, as it may happen, for instance, with lightning arresters.] This half-cycle interruption is an extreme case of notching as, for instance, caused by SCR switching, which also is eliminated satisfactorily only with version α.

4.5.3.2 Limiters. If transient spikes are additive to and exceeding the extremes of the useful voltage (and not subtractive like notches, which are most economically handled by the means outlined in 4.5.2), limiting elements, so-called nonlinear filters, are indicated.

For signal lines, back-to-back diodes (avalanche or selenium), shunted across the lines, are good protectors, for instance, for differential amplifier inputs. Current impulses are drastically reduced by positive temperature coefficient resistors put in series with the line.

For lines carrying power, highly energetic spikes are squelched by air gap suppressors, sharp kneed selenium rectifiers (back to back for alternating

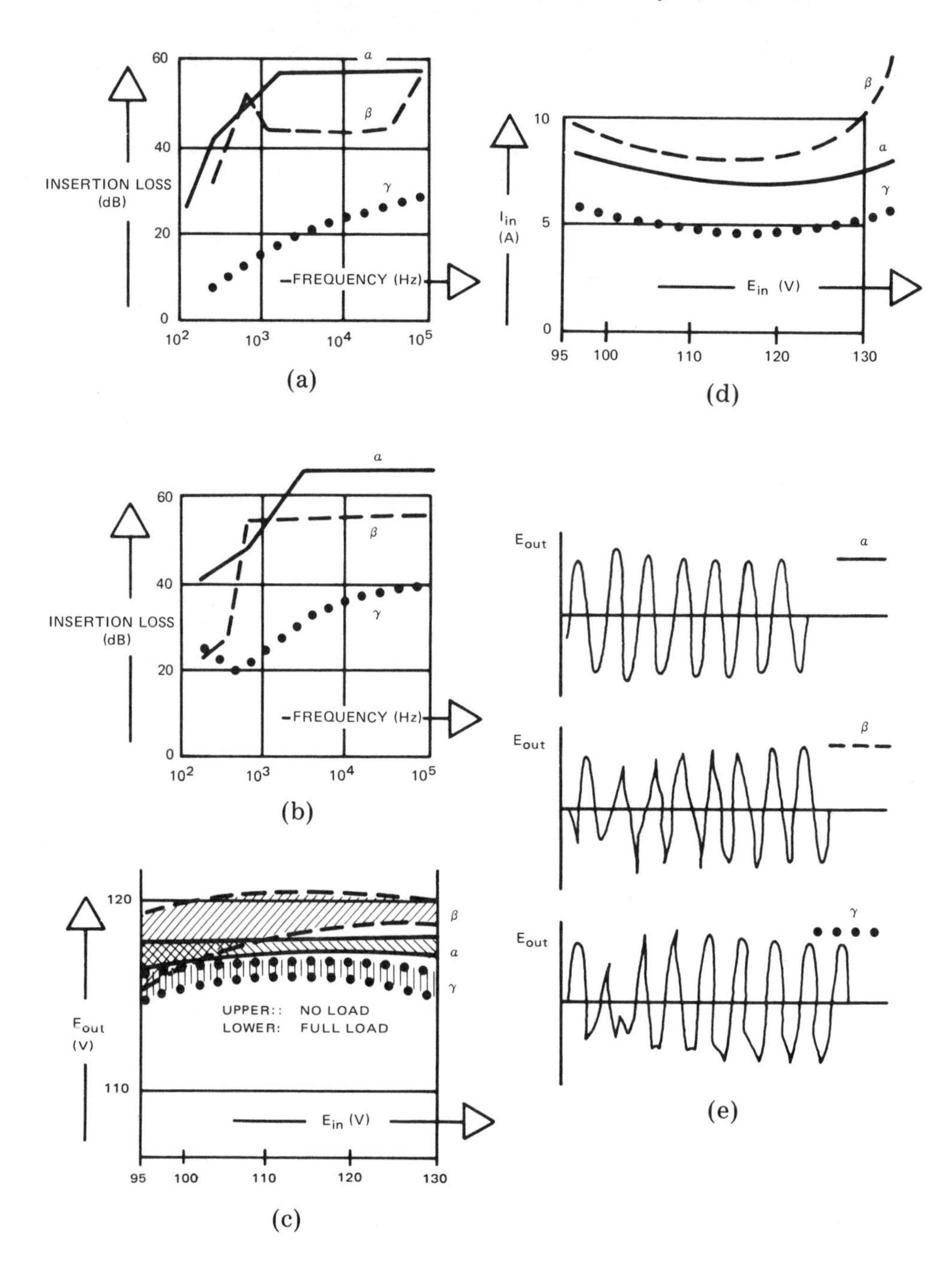

Fig 57
Juxtaposition of Key Data of Three Magnetic-Core Filter Regulators
(a) Forward Insertion Loss. (b) Reverse Insertion Loss. (c) Regulation.
(d) Input Current for Full Load. (e) Response to Half Cycle Interruption

current and very rugged), or zinc-oxide nonlinear resistors, so-called varistors. Their selection depends on the expected rating of the spikes. In digital systems clipping (by emitters) can render the spikes harmless if and only if the spike width is much smaller than the width of the digit.

4.5.3.3 Information Matched Filters. This section is briefly concerned with noise in signal lines, specifically with signals which are buried seemingly irretrievably in noise, at least not retrievable with conventional S-domain filters, irrespective of how much effort is being made.

It is beyond the scope of this guide to elaborate in any detail on the methods to extract such signals from noise, yet the reader should at least be made aware that such sophisticated filters exist.

To remove signals from completely masking noise, with signal and noise power spectra completely overlapping, one exploits, expressed very simply, the fact that noise if averaged in time or for different ensembles or if correlated (compared at different times) cancels out to zero (assuming white noise). The signal, in contrast, if averaged or correlated (with itself—autocorrelation—or with a coperiodic other signal—cross correlation) either reveals its original shape (averaging) or at least its presence and period (missing phrase relations change signal appearance in correlation analysis).

For digital signals, matched filters are matched to the signal form and respond essentially only to the signal for which the filter is matched, telling whether or not the signal is there.

Whenever something is known about the signal, phase-locked loops can detect signals 30–40 dB below the noise level. And if one takes sufficient time, averaging or correlating filters can extract signals from noise of unbelievably poor signal-to-noise ratio.

Just to give an example, Fig 58 shows a signal completely immerged in noise. After averaging a thousand consecutive samples (prior knowledge of signal periodicity is required), the extracted signal becomes quite obvious, and essentially perfect with an increase in time.

Impulsive noise in digital systems (with the exception of cases discussed before) is difficult to suppress once it has reached critical signal or control lines. Hence coupling into critical lines must be avoided.

If operation in a high-noise environment is to be expected, it is best to select from the very beginning a system into which "filtering" has been designed as a systems mode. Typical examples are the incorporation of redundancy, error-correcting codes, etc, which serve to minimize the probability of error.

4.5.3.4 Directional or Mode Filters. In lines (wire pairs) one normally talks about a forward and a return wire (normal mode). Quite often, however, one encounters common-mode propagation, mostly caused by multiple grounds, and consisting of noise current flow in the same direction for both wires. Figure 59 depicts the common mode as contrasted to the normal

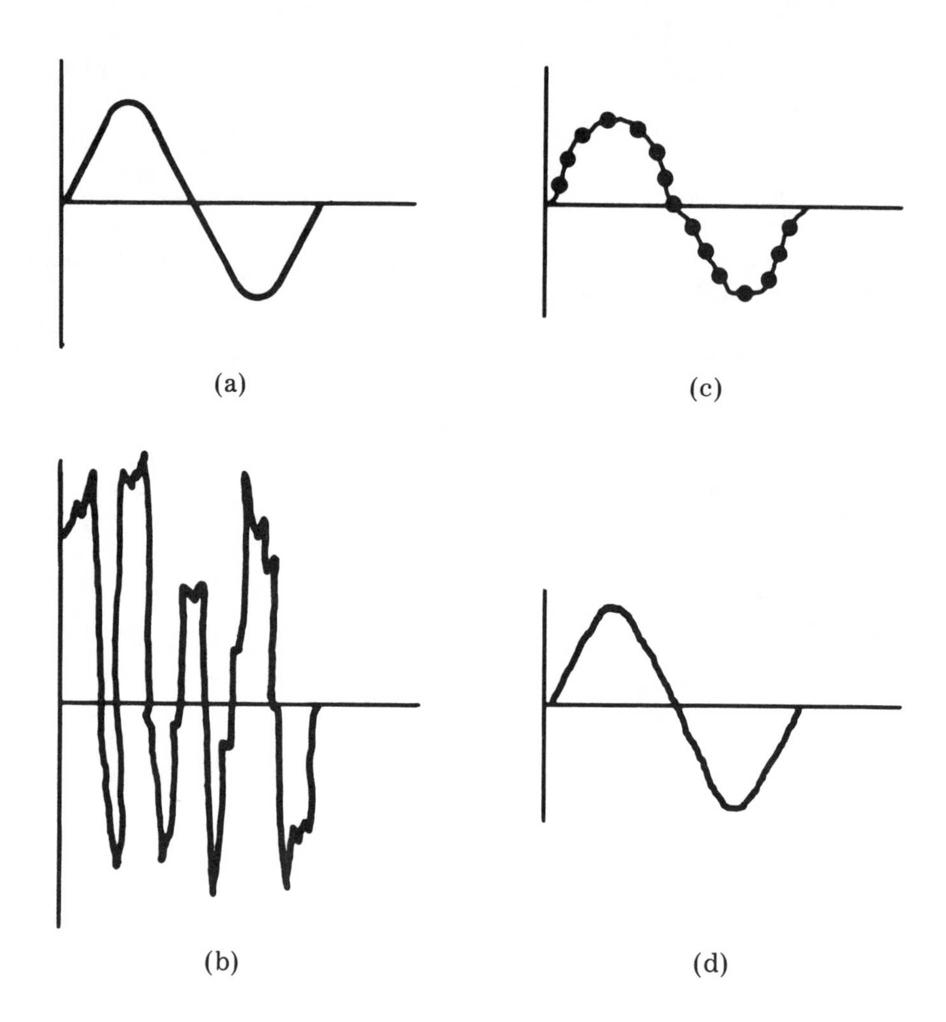

Fig 58
Extracting Signal by Averaging. (a) Original Signal.
(b) Instant Signal + Noise. (c) Short-Time Averaging.
(d) Long-Time Averaging

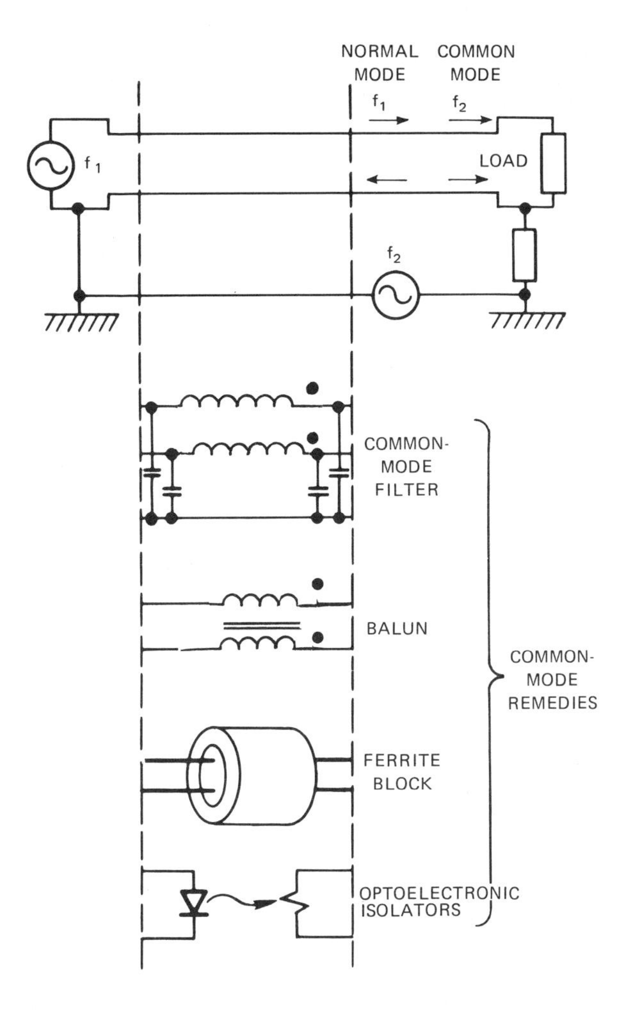

Fig 59
Common-Mode Operation and Some Remedies

(Not shown are isolation transformers and high-common-mode-rejection operational amplifiers. In this context we refer to Fig 57(a) and (b), which shows the normal-mode insertion loss of isolation transformers, the common-mode rejection of which is extremely good, particularly at low frequencies.)

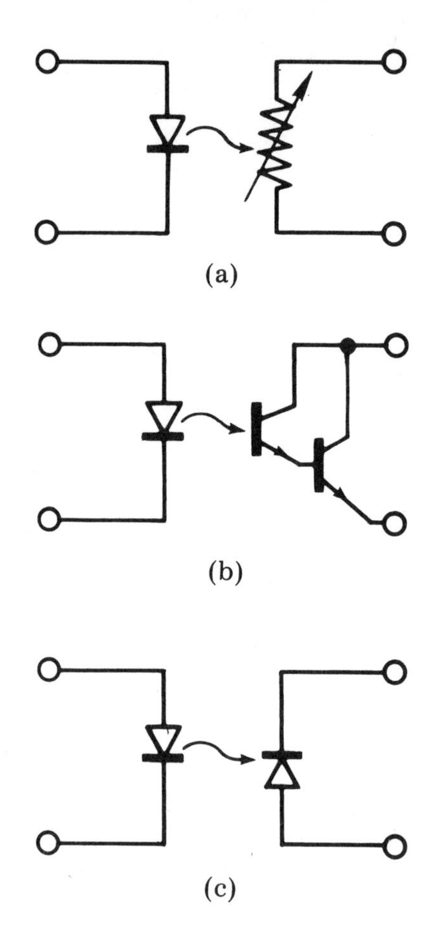

Fig 60
LED-Operated Electrooptical Isolators
(a) CdS Photoconductor, on 3 ms, off 0.1 s; Is Simultaneously Low-Pass Filter. (b) Si Photo Darlington, on 10 μs, off 40 μs. (c) Si Photodiode, 5 ns Switching

mode and also some effective remedies to suppress the common-mode propagation.

Since the current flow provided by the normal mode does not create any biasing effects for the inductors, the common-mode filters, baluns, or ferrite slugs do not become saturated and hence, can be held quite small.

Optoelectric isolators (Fig 60) quite often use the light source light-emitting diodes for long life (in contrast to lamps). Three basic light receivers are available. For low speeds CdS sensors are most desirable because they also eliminate transients of the normal mode [Fig 60(a)]. For higher speeds the versions shown in Fig 60(b) and (c) using silicon receivers are indicated. Linearity, which is important in signal lines, has to be checked.

4.6 References

[1] ANSI C63.2-1980, Specifications for Electromagnetic Noise and Field-Strength Instrumentation, 10 kHz to 1 GHz.

[2] ANSI C63.4-1981, Methods of Measurement of Radio-Noise Emissions from Low-Voltage Electrical and Electronic Equipment in the Range of 10 kHz to 1 GHz.

[3] ANSI C63.12, Draft American National Standard, Recommended Practice on Procedures for Control of System Electromagnetic Compatibility.

[4] ANSI/IEEE C37.90-1978, IEEE Standard for Relays and Relay Systems Associated with Electric Power Apparatus. (Includes ANSI/IEEE Std 472-1974 (R 1979), Guide for Surge Withstand Capability (SWC) Tests.

[5] ANSI/IEEE C57.98-1968, Guide for Transformer Impulse Tests (appendix to ANSI/IEEE C57.12.90).

[6] ANSI/IEEE Std 4-1978, IEEE Standard for High-Voltage Testing Techniques.

[7] ANSI/IEEE Std 213-1961 (R 1974), Methods of Measurement of Radio Interference: Conducted Interference Output to the Power Line from FM and Television Broadcast Receivers in the Range of 300 kHz to 25 MHz.

[8] ANSI/IEEE Std 430-1976, Procedures for Measurement of Radio Noise from Overhead Power Lines.

[9] ANSI/NEMA ICS Package, Standards on Industrial Controls and Systems.

[10] ANSI/NFPA 70-1981, National Electrical Code.

[11] ANSI/NFPA 79-1980, Electrical Standard for Metalworking Machine Tools and Plastics Machinery.

[12] IEC Publ 60-1 to 60-4, 1977, High-Voltage Test Techniques, Parts 1-4.

[13] IEEE Std 139-1952, IEEE Recommended Practice for Measurement of Field Intensity above 300 MHz from Radio-Frequency Industrial, Scientific, and Medical Equipment.

[14] IEEE Std 284-1968, IEEE Standards Report on State-of-the-Art of Measuring Field Strength, Continuous Wave, Sinusoidal.

[15] IEEE Std 748-1979, Standard for Spectrum Analyzers.

[16] JIC Standards and Tables for Machine Tool and Mass Production Equipment. *Electrotechnology*, Section E11, Apr/May 1967.

[17] MIL Std 220A.

[18] MIL Std 461.

[19] MIL Std 461A/462.

[20] MIL Std 826A.

[21] NASA Spec 279. Marshall Space Flight Center, Houston, TX.

[22] BECK, E. *Lightning Protection of Electric Systems*. New York: McGraw-Hill Book Co, 1954.

[23] BENNISON, E., GHAZI, A.J., and FERLAND, P. Lightning Surges in Open Wire, Coaxial, and Paired Cables. *IEEE Transactions on Communications*, vol COM-21, Oct 1973, pp 1136-1143.

[24] BODLE, D.W., and GRESH, P.A. Lightning Surges in Paired Telephone Cable Facilities. *Bell System Technical Journal*, vol 40, Mar 1961, pp 547-576.

[25] BROAD, W.J. Nuclear Pulse (I): Awakening to the Chaos Factor. *Science*, vol 212, May 29, 1981, pp 1009-1012.

[26] CHOWDHURI, P. Breakdown of *p-n* Junctions by Transient Voltages. *Proceedings of the National Electronics Conference*, vol 20, 1964, pp 340-345: also *Direct Current*, vol 10, Aug 1965, pp 131-139.

[27] CHOWDHURI, P. Transient-Voltage Characteristics of Silicon Power Rectifiers. *IEEE Transactions on Industry Applications*, vol IA-9, Sept/Oct 1973, pp 582-592.

[28] CHOWDHURI, P. Transient Voltages in Transit Systems. *IEEE Transactions on Electromagnetic Compatibility*, vol EMC-17, Aug 1975, pp 140-149.

[29] CHOWDHURI, P. Breakdown of Thyristors under Transient Voltages. *Proceedings of the 1980 IEEE PES Winter Meeting*, IEEE Pub 80 CH1523-0 PWR, 1980, Paper A80 114-9.

[30] CHOWDHURI, P., and BOCK, D.D. Coordination of Transient-Voltage Characteristics of Electrical Systems. *IEEE Transactions on Industry Applications*, vol IA-9, Sept/Oct 1973, pp 577-581.

[31] CHOWDHURI, P., and WILLIAMSON, D.F. Electrical Interference from Thyristor-Controlled DC Propulsion System of a Transit Car. *IEEE Transactions on Industry Applications*, vol IA-13, Nov/Dec 1977, pp 539-550.

[32] CHOWDHURI, P., and ZOBRIST, D.W. Susceptibility of Electrical Control Systems to Electromagnetic Disturbances. *IEEE Transactions on Industry Applications*, Vol IA-9, Sept/Oct 1973, pp 570-576.

[33] CLARK, D.B., BENNING, R.D., KERSTEN, P.R., and CHAFFEE, D.L. Power Filter Insertion Loss Evaluated in Operatonal-Type Circuits.

IEEE Transactions on Electromagnetic Compatibility, vol EMC-10, June 1969, pp 243-255.

[34] FISCHER, D.F., and COWDELL, R.B. New Dimensions in Measuring Filter Insertion Loss. *EDN*, May 1969.

[35] GAUPER, H.A., HARNDEN, J.D., and McQUARRIE, A.M. Power Supply Aspects of Semiconductor Equipment. *IEEE Spectrum*, Oct 1971, pp 32-43.

[36] GEISHEIMER, F. Recommendations of an Engineering Contractor. *IEEE Transactions on Industry and General Applications*, vol IGA-3, Mar/Apr 1967, pp. 83-87.

[37] GENTRY, F.E., *et al. Semiconductor Controlled Rectifiers*. Englewood Cliffs, NJ: Prentice-Hall, 1964.

[38] GLASSTONE, S. (Ed). *The Effects of Nuclear Weapons*, US Atomic Energy Commission, Apr 1962, Chap 10.

[39] GOERS, R.E. Quiet Wiring Zone. *IEEE Transactions on Industry and General Applications*, vol IGA-5, May/June 1969, pp 273-277.

[40] GOLDE, R.H. *Lightning Protection*. London: Edward Arnold, 1973.

[41] GOLDE, R.H. (Ed). *Lightning*, vols 1 and 2. New York: Academic Press, 1977.

[42] GOODING, F.H., and SLADE, H.B. Shielding of Communcation Cables. *AIEE Transactions (Communication and Electronics)*, vol 74, July 1955, pp 378-387.

[43] HAYS, J.B. Protecting Communication Systems from EMP Effects of Nuclear Explosions. *IEEE Spectrum*, May 1964, pp 115-122.

[44] HOLM, R. *Electric Contacts*, 4th ed. New York: Springer-Verlag, pp 272, 283, 289.

[45] HUELSMAN, L.P. *Theory and Design of Active RC Networks*. New York: McGraw-Hill Book Co, 1968.

[46] Joint Special Issue on the Nuclear Electromagnetic Pulse. *IEEE Transaction on Electromagnetic Compatibility*, vol EMC-20, Feb 1978.

[47] KLIPEC, B.E. Reducing Electrical Noise in Instrument Circuits. *IEEE Transactions on Industry and General Applications*, vol IGA-3, Mar/Apr 1967.

[48] KOERITZ, K.W., and ROBSON, C.A. A Systems and Environmental EMC Control Program for the AIRTRANS Automated Ground Transportation System. *1974 IEEE Electromagnetic Compatibility Symposium Record*, IEEE Pub 74CH0803-7 EMC, 1974, pp 189-197.

[49] LERNER, E.J. EMPs and Nuclear Power. *IEEE Spectrum*, vol 18, June 1981, pp 48-49.

[50] MAHER, J.R. Cable Design and Practice for Computer Installation. Presented at the 19th Engineering Conference of TAPI, Seattle, WA, July 29-30, 1964; Publ of Tappi Technical Association of the Pulp & Paper Industry, vol 48, May 1965, pp 90A-93A.

[51] MILLS, G.W. The Mechanisms of the Showering Arc. *IEEE Transactions on Parts, Materials, and Packaging*, vol PMP-5, Mar 1969, pp 47-55.

[52] RALOFF, J. EMP—A Sleeping Electronic Dragon. *Science News*, vol 119, May 9, 1981, pp 300-302.

[53] RALOFF, J. EMP Defensive Strategies. *Science News*, vol 119, May 16, 1981, pp 314, 315.

[54] RUDENBERG, R. *Electrical Shock Waves in Power Systems*. Cambridge, MA: Harvard University Press.

[55] SCHENKEL, H., and STATZ, H. Voltage Punch-Through and Avalanche Breakdown and Their Effect on the Maximum Operating Voltages for Junction Transistors. *Proceedings of the National Electronics Conference*, vol 10, 1954, pp 614-625.

[56] SCHLICKE, H.M., *Electromagnetic Compossibility*. (Applied Principles of Cost-Effective Control of Electromagnetic Interference and Hazards), 2nd Enlarged Ed. New York: Marcel Dekker, 1982.

[57] SCHLICKE, H.M., BINGENHEIMER, A.J., and DUDLEY, H.S. Elimination of Conducted Interference, A Survey of Economic, Practical Methods. Invited paper at the 1971 Annual Meeting of the IEEE Industry and General Applications Group.

[58] SCHLICKE, H.M., and WEIDMAN, H. Compatible EMI Filters. *IEEE Spectrum*, vol 4, Oct 1967, pp 59-68.

[59] SCHLICKE, H.M., and WEIDMAN, H. Effectiveness of Interference Filters in Machine Tool Control. Presented at the 19th Annual IEEE Machine Tool Conference, Detroit, MI, Oct 1969.

[60] SCHULZ, R.B., PLANTZ, V.C., and BRUSH, D.R. Shielding Theory and Practice. Presented at the 9th Tri-Service Conference on EMC, 1963.

[61] SCHWAB, A.J. *High-Voltage Measurement Techniques*. Cambridge, MA: MIT Press, 1972.

[62] Shielding for Electromagnetic Compatibility. Magnetic Metals Company, NJ.

[63] SKOMAL, E.N. Distribution and Frequency Dependence of Unintentionally Generated Man-Made VHF/UHF Noise in Metropolitan Areas. *IEEE*

Transactions on Electromagnetic Compatibility, vol EMC-7, 1965, pp 263-277.

[64] SKOMAL, E.N. Comparative Radio Noise Levels of Transmission Lines, Automotive Traffic, and RF Stabilized Arc Welders. *IEEE Transactions on Electromagnetic Compatibility*, vol EMC-9, 1967, pp 73-77.

[65] STACEY, E.M., and SELCHAU-HANSEN, P.U. SCR Drives — AC Line Disturbances, Isolation, and Short-Circuit Protection. *IEEE Transactions on Industry Applications*, vol IA-10, Jan/Feb 1974, pp 88-105.

[66] SUNDE, E.D. *Earth Conduction Effects in Transmission Systems.* New York: Dover Publications, 1968.

[67] SWINEHART, M.R. Electrical Noise in Machine Tool Controls. *IEEE Transactions on Industry Applications*, vol IA-8, Sept/Oct 1972, pp 535-541.

[68] UMAN, M.A. *Lightning.* New York: McGraw-Hill Book Co, 1969.

[69] VASAKA, C.S. Problems in Shielding Electronic Equipment. *Proceedings of the Conference on Radio Interference Reduction*, 1954.

[70] WAGAR, H.N. Prediction of Voltage Surges when Switching Contacts Interrupt Inductive Loads. *IEEE Transactions on Parts, Materials and Packaging*, vol PMP-5, Dec 1969, pp 149-155.

[71] WAVRE, A. Application of Integrated Circuits to Industrial Control System with High Noise Environments. *IEEE Transactions on Industry and General Applications*, vol IGA-5, May/June 1969, pp 278-281.

[72] WHITE, D.R.J. *EM: Test Instrumentation and Systems*, vol 4. Gaithersburg, MD: Don White Consultants.

[73] YOUNG, F.J. Pulse Shielding by Nonferromagnetic and Ferromagnetic Materials. *Proceedings of the IEEE*, Vol 61, Apr 1973 pp 404-413.

5. Systems Approach to Noise Reduction

5.1 Introduction. Since this guide is primarily concerned with minimizing noise effects by installation, hardening of the control system itself against noise is, by definition, outside the scope of this guide. Rather a noise-hardened system is tacitly assumed from the very beginning. However, since this assumption is not always valid, a set of brief statements pertaining to design *and* testing for noise hardness seems to be in order. Even the most diligent, costly, and circumspect noise-immune installation of a system is not enough if the system itself is highly susceptible and so unstable that it responds to the slightest noise.

5.1.1 Problems in Microprocessor Systems. The control of electromagnetic interference has come a long way from the early retroactive radio-frequency interference "fixes" to the predictive electromagnetic compatibility planning thoroughly applied to modern military and communications systems.

If one tries to apply such well-established electromagnetic-compatibility methods to computer or microprocessor-dominated industrial control systems, or both, one becomes quickly disappointed. Many aspects of conventional electromagnetic-compatibility methods are just not applicable to such interference-prone systems. There are several reasons for this.

(1) Much of electromagnetic-compatibility is essentially concerned with communications systems. In these, sources of electromagnetic interferences are essentially intentional, narrowband transmitters, the channels of transmission are clearly defined, and the receivers are tuned to the intentional transmitters. In contrast, in modern industrial control systems the sources of interference are mostly unintentional broadband transients or field concentrations. The transfers of interference are ubiquitously spread throughout the whole system, whereby a multiplicity of often unsuspected receptors is unavoidably immersed into a very noisy space in which sensing and control of the processes and machines takes place.

In modern control systems we operate under disparate conditions requiring an approach distinctly different from the well-established military communications-related electromagnetic-compatibility methods. We shall identify it as electromagnetic compossibility to indicate an idiosyncratic, previously neglected, branch of the generic concept of electromagnetic-compatibility. (For details see [2].)

(2) One cannot expect to convert a complex problem into independent simple subproblems without much thinking. In other words, the "cookbook" engineer who wants to follow a table of simple precepts will not be able to control electromagnetic interference cost-effectively in intrinsically nonlinear, intraactive systems, unless the system and electromagnetic compatibility are coplanned.

(3) Unfortunately engineering education is not particularly conducive to good electromagnetic-compatibility thinking: (a) Electromagnetic compatibility, requiring truly interdisciplinary thinking, is not taught in schools.

(b) Engineering education teaches engineering laws, not adaptive engineering thinking. (c) The recent emphasis on digital techniques in favor of analog and distributed-parameter techniques somehow limits an electromagnetic-compatibility-commensurate understanding of many engineers.

(4) The attitude of industrial management is changing rapidly from disregarding electromagnetic compatibility to mandating it. The long-range costs, directly and indirectly, of the nonelectromagnetic-compatibility-planned design of systems are becoming more and more recognized.

5.1.2 Objectives. It is unnecessary to rewrite the many standards that comprise the methodology of established electromagnetic compatibility. Many of these are fine as long as one realizes that they are addressing an isolated problem. But often an electromagnetic-interference problem is so much oversimplified that the remedy creates a counterpositive effect. Consequently, simplistic attempts to make things tractable may actually make them intractable.

The objectives of this section are very much condensed from [2] [3] and are as follows: (1) To develop a systemic approach of electromagnetic compossibility for industrial and commercial systems, such that contrary objectives and principles can be made to coexist by proper planning. (2) To stress principles, not simplistic precepts, recast theory for practical application, and take away the stigma of having to rely on trial-and-error methods. (3) To amend ingrained misconceptions of such basic electromagnetic-compatibility measures as grounding, shielding, and filtering, which are supposedly very simple, but must be considered as part of a complex system.

5.2 The Systemic Approach

5.2.1 Analyzing the System. In order to come to grips with a complex situation, one has to start with an analysis, first of the individual electromagnetic-interference links and their components and then of their interactions. Many critical parameters are involved. The easily remembered acronym FATTMESS should be applied to each of the often unsuspected sources, transfers, and receptors of interference. FATTMESS stands for frequency, amplitude, time, temperature, mode, energy, size or structure, and statistics. Not all criteria are significant for each entity, but should be checked so that nothing is overlooked.

Because of the heterogeneity of objectives and conditions, there must be a continual feedback and feedforward between analysis and control for optimization and cost effectiveness.

In the initial electromagnetic-interference analysis of the system as a whole, we must accept five conditions imparted by the nature of an industrial control system.

(1) Because the system under consideration is a control system, elements of the system are highly interactive.

[3] The numbers in brackets correspond to those of the references listed in 5.4.

(2) This interaction is further amplified by mandatory safety measures ANSI/NFPA, in particular as they pertain to grounding, bonding, and protection devices. Line frequency and transients are diffused untransparently into the system. They include dangerous secondary transients or field gradients caused by the very devices employed for safety.

(3) Quasi-static considerations expressed by the term *control common* are inappropriate for large systems, even for those lines supposedly carrying only low-frequency signals. Many transients contain high-frequency components requiring a transmission-line approach. It is important to remember that a quarter wave length of a line converts a short circuit into an open circuit and vice versa. Also, radio frequency is rectified in transistors, thus changing their operating conditions. On the basis of points (2) and (3) no two metalically connected points can be considered as being at the same potential. Rather, they must be viewed as an interference source with a source impedance (although it is possible to create quasi-grounds in limited areas, see [2], Ch 5).

(4) Normal-mode conversion (normal mode to common mode) and reconversion is common as wiring does not constitute the nearly ideal transmission lines applied in communciations systems.

(5) The sensors and activators of the control systems are by necessity *exposed* receptors, which must work in a noisy space caused by the switching of power in the system or a near-neighbor system.

5.2.2 The Ideal System. After the brief foregoing initial systems analysis, the first and crucial stage of systemic control may be approached with the proper perspective. It is necessary to order the electromagnetic compatibility disorder, reconcile the seemingly irreconcilable. One cannot violate the NEC and one cannot have a rigid mind going strictly by the book only. Retroactive introduction of electromagnetic-compatibility measures can be a very costly affair (Fig 61). Three key steps are involved in this adaptive coplanning process:

(1) Partition of system

(a) Quiet spaces. This contains low-amplitude fast logic. Logic no faster than necessary.

(b) Semiquiet spaces. The interfaces contain slow high-amplitude input-output devices. Make the speed no greater than necessary.

(c) Unavoidably noisy spaces. This contains in addition to power devices, the exposed receptors.

The partitioning is well facilitated by microprocessors which permit establishing distributed subsystems, which again can be partitioned as given above.

(2) Isolation

(a) By shielding and filtering of quiet and semiquiet spaces

(b) Of actuators (control elements) by electrooptical isolators

(c) Of exposed receptors of great susceptibility by using the following:

(i) Noncontacting sensors or electrooptical isolators

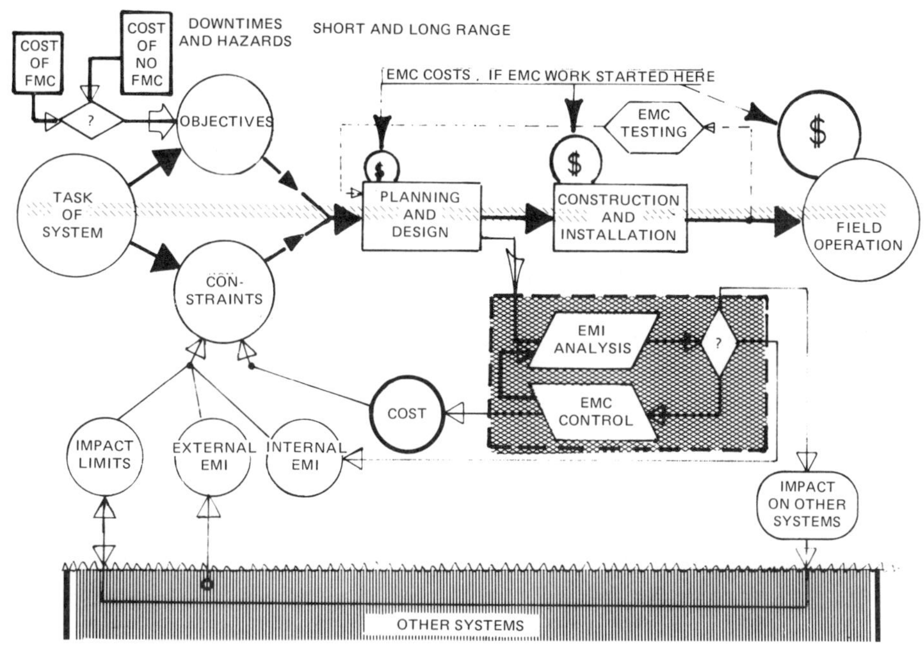

Fig 61
System and Electromagnetic Compatibility Planning

(ii) Isolating amplifiers with isolation for input-output *and* power supply

(iii) Fiber optics for critical exposed transfers. The radiative coupling into such transfers is absolutely zero. But naturally, there are still the electromagnetic-interference problems of normal-mode errors (or failures) affecting the electrooptical transducers themselves. It is still necessary to resort to conventional electromagnetic-compatibility measures to immunize the electrooptical transducers. Nevertheless we can be sure that lightning and line frequency ground faults, and even ground stray currents, find transfer $T = 0$ in fiber optics.

The electrooptical isolators and fiber optics are rapidly evolving hardwares. Their manufacturers often supply detailed application notes on their products. Thus we can confine ourselves to principles and pitfalls involved.

(3) Redundancy. Redundancy of data-handling subsystems is now economically feasible. Redundancy may be either duplication or triplication of whole critical subsystems, or may be error-correcting coding. In [2] the bibliography lists on p 167 excellent tutorial papers on this important subject. Thus error and fault tolerance can be programmed into the system by redundancy of the data-handling subsystems and by software. For the power-handling subsystems, redundancy would be too expensive. The redundancy must be replaced by subdividing power aggregates, such that emergency operation is still feasible when one of the power aggregates fails.

In this generic first step of systemic electromagnetic-compatibility assurance planning significant developments, still very much in flux should be considered.

(1) Electrooptics. Isolatiors replace wires for information handling. For this we have *sources:* light-emitting diodes (LED) or lasers, *modulators:* crystals of lithium niobate or tantalate, *receptors:* photodiodes and phototransistors, and *transfers:* electrooptical isolators eliminating the effects of double grounding and common-mode coupling; they can be made linear; and fiber optics eliminating all undesirable transfer and coupling.

(2) Microprocessors. Extensive (programmable) wiring of logic at extremely low cost, in very small space, with quite high reliability. Distinquish between (i) 1 bit machines (replace wired logic, decision oriented) and (ii) multiple-bit machines (data logging). Networks (either bus or memory-oriented) of (i) or (ii), or (i) and (ii) approach computer capabilities. Hardware is replaced by software.

(3) Charge-Coupled Devices. They are actually shift registers for analog charge signals, whereby the charges they store and transfer are introduced electrically or optically, free of switching transients. Charge-coupled devices need no analog-to-digital and digital-to-analog conversion. Great reduction in digital hardware results, with very fast scanning rates of linear and planar image sensors (much faster than vidicons). These are sophisticated, noncontacting, self-scanning sensors which can also be used for signal correlating, signal reformulating, multiplexing, and digital filtering.

(4) Solar cells. Solar cells with a practical efficiency of about 10% are now available, such that the power supply of preamplifiers in extremely hazardous situations can now be made wireless, thus harmless.

Thus in principle we are now able to render the spread transfer harmless.

5.2.3 Flexibility in, and Planning of the Near-Ideal System. At present there are still some constraints to build ideally interference-free systems. The price, the aging, and the limited linearity and dynamic range of optoelectronic devices may, at present, necessitate a judicious application of more conventional electromagnetic-compatibility measures.

Besides localized suppression at the source, in transfer, or at the receptor, and besides passive and active cancellation techniques, there are many options for differentiation between wanted and unwanted signals.

In the process of differentiation, electromagnetic-compatibility measures should be viewed from as many angles as possible, such that the problem appears—from one of these angles—in its greatest simplicity. Then the system with a minimum of cost can be firmed up. Naturally the various options should be assessed carefully for contingencies.

In its simplest form, differentiation is direct. This means staying with the key issue of our FATTMESS criterion. For instance, stay within the frequency domain by using a low-pass filter if the wanted signal is low-frequency F and the interfering source is high-frequency F. Thus such direct differentiation is, for example, symbolized by F/F and the enclosing circle. But quite often the distinction within one criterion of FATTMESS is rather expensive, blurred, or difficult to view of certain concomitants. Then it is quite effective to resort to multiple-criteria or indirect differentiation. See the lower part of Fig 62, where different kinds of lines mark the various criteria involved. For instance, column (a) refers to amplitude-frequency conversion having much the great advantage that FM has over AM. Column (b) pertains to multiple-realm timing effects. Column (c) handles notching (short spikes within the envelope of low frequency) in power lines. For example, compare three options available to get rid of notching carried on signal lines. It seems that the amplitude criterion A is the key issue. Obviously conventional clippers or limiters of A cannot be used. But limiting can be done indirectly, for instance by referring to (iii) and (iv) of column (b). Figure 63 juxtaposes a slew-rate limiter (exploiting frequency F, time T, and energy E) and two rise-time limiters (exploiting a basic realtionship between frequency F and time T). Systemic requirements determine which of the three options is most appropriate in a specific case.

5.2.4 Updating Outdated Oversimplifications. It is often assumed that the long-established electromagnetic-compatibility concepts of grounding/bonding, filtering, and shielding are independent modes of action because they seem to be simple. But quite often, one hears the angry remark, "it does not work." The reason again is, that the published prescriptions of these electromagnetic-compatibility means are based on oversimplified premises taken out of the systemic context. Conditions become more complex as the more modern control systems are getting more sophisticated and larger. The misconceptions (and their corrections) that plague these weak cornerstones of conventional electromagnetic compatibility are noted in 5.2.4.1.

5.2.4.1 Wiring/Grounding. As previously stated the concept of control common—simulating a treelike arrangement of the signal return wires—loses its meaning in large systems. ANSI/IEEE Std 100-1977 [1], has 95 entries for grounding—not of much help to the engineer concerned with specific multiple grounding problems. It makes no sense to give the ground a special name or symbol as long as there are interactions. Such wishful designation does not impart the specified ground function only. The only reasonable way out of this dilemma is to make part of the wiring into nonwiring by isolation.

5.2.4.2 Powerline Filtering. Powerline filtering is indispensable, but it is still a highly controversial topic because it is still based on simplistic engineering standards. When more than 50 years ago, the admirable theory of LC filters was developed it clearly stated its premise, impedance match. But this very decisive premise does not at all hold for power feed line filters of large systems for the following reasons:

(1) Severe indeterminate mismatch is unavoidable and may result in poor filtering low in the stopband and/or insertion gain (up to 40 dB in the upper passband). The 3 dB point of conventional filter theory loses all meaning as it may be 3 dB plus or minus 40 dB.

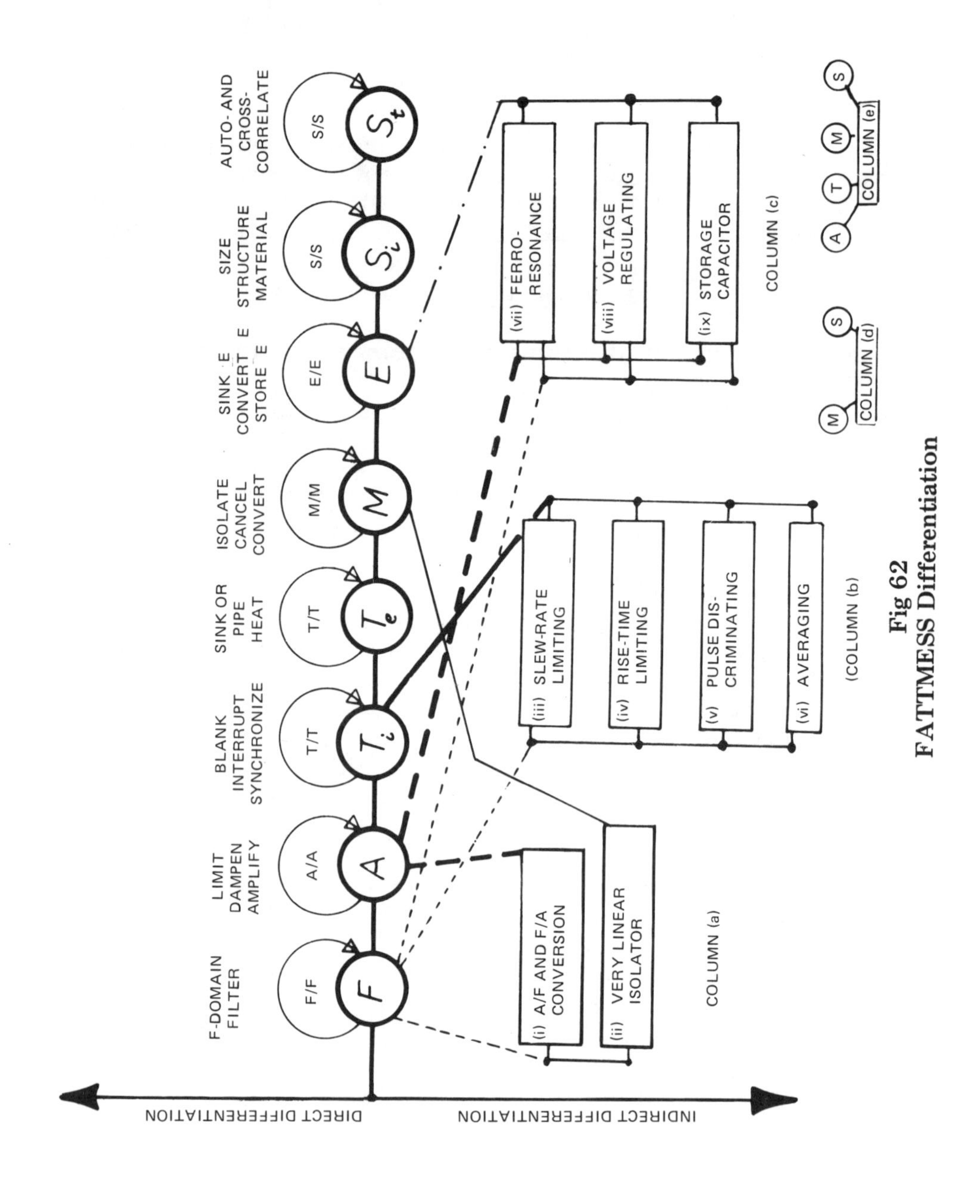

Fig 62
FATTMESS Differentiation

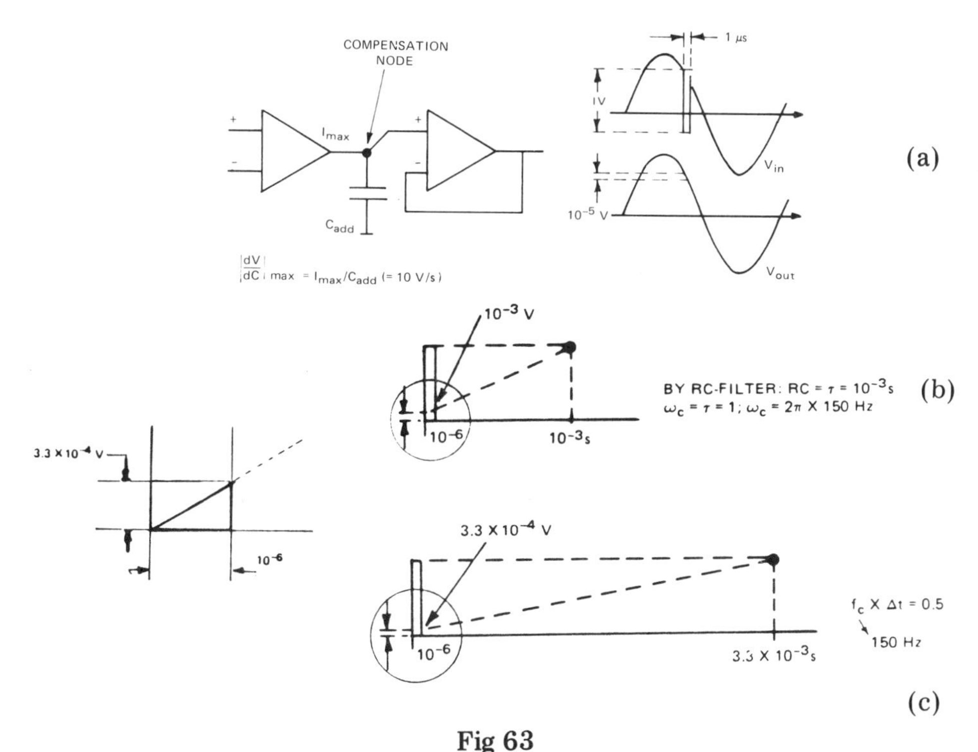

Fig 63
Filtering of Notches in Single Lines.
(a) Slew-Rate Limiting. (b) Risetime Limiting by RC Filter.
(c) Risetime Limiting by Sharp Cutoff Filter

(2) Switching, cyclically or randomly, or loads and sources creates highly detrimental ringing, doubling the switched amplitude and causing decaying oscillations.

(3) Brute-force overdesign of filters, by providing a specious safety margin for mismatch, is costly and worthless

(4) It's not practical to rely on filter measurements made in the conventional uncritical 50 Ω/50 Ω interface condition, but applied to the real world of large, switched systems.

There is a simple and economic way out of the seemingly perplexing situation. Design low-pass filters as multiple-section filters and make them lossy in the upper passband. Acceptance test them in 0.1/100 and 100/0.1 Ω interfaces for the frequency range of 2—150 kHz (or 15 kHz). If the filter meets the requirements for those conditions, all other filter testing can be done in the 50 Ω/50 Ω system: the type testing above 150 kHz and all quality-control testing over the whole frequency range. This realistic worst case method is now under consideration by CISPR. (See also [2], Chs 8 and 11.)

5.2.4.3 Shielding. To be more realistic, the conventional plane-wave approach has to be discarded. Size and structure are important parameters. These critical conditions have been taken into consideration by the universal shielding diagrams given in 4.4.3. (See [2], also with respect to the detrimental effect of biasing.)

5.3 Noise Immunity Test. To be sure that the control system is properly designed, one has to test it for noise *immunity*. This is still in an embryonic state due to the lack of a comprehensive data base of noises encountered in industrial control environments. At present there are three significant tests which can at least serve as indicators of noise immunity. They will be developed into true tests with clearly defined limits once the noise classification is further advanced.

5.3.1 Relay Noise Test. This test tends to simulate unsuppressed relays operating off the same power line and located in close proximity to solid-state control systems. The test circuit shown in Fig 64 consists of two or three industrial-type control relays which interrupt their own coil circuit. Electrical connections are made to the relays through timing contacts to prevent overheating and destruction of the coils. Noise inducer coils are formed and wired in series with the coil circuit. Interruption of the coil circuit causes high-voltage spikes at the inducer coils, which are then capacitively coupled into the control system.

With this method it is possible to determine sensitive areas of the system. The test also permits testing of the shielding of the sensitive portion of the control circuit. It can be shown that proper shielding will eliminate most if not all of this type of noise interference. This test is intended to simulate a *worst case* industrial noise environment. The *worst case* is often actually the normal case.

5.3.2 Motor Noise Test. This test is conducted to establish some qualitative analysis of the behavior of the control system against a high-power electromagnetic field of predominantly fundamental frequency and tends to simulate the presence of heavy bus bars feeding motors with power ratings of several hundred kilowatts. The electromagnetic field is simulated by switching squirrel-cage motors from full forward to full reverse (Fig 65). One phase of the three-phase motor supply leads will be used to set up the magnetic field which will permeate the solid-state control circuit. Depending on the magnitude of the current surges, the number of turns is selected to adjust the field strength. Loops are applied in horizontal and vertical directions and may be used alternately.

Figure 66 shows a typical motor test with the magnetizing windings in the vertical direction generating the horizontal field, and the horizontal magnetizing windings generating the vertical field. The windings can be applied selectively and thus allow the direction of the electromagnetic field to be adjusted for maximum noise coupling. To the test unit a total of 2000 A has been applied without affecting the operation of this control system at an ambient temperature of 70 °C. As the temperature is elevated, the switching

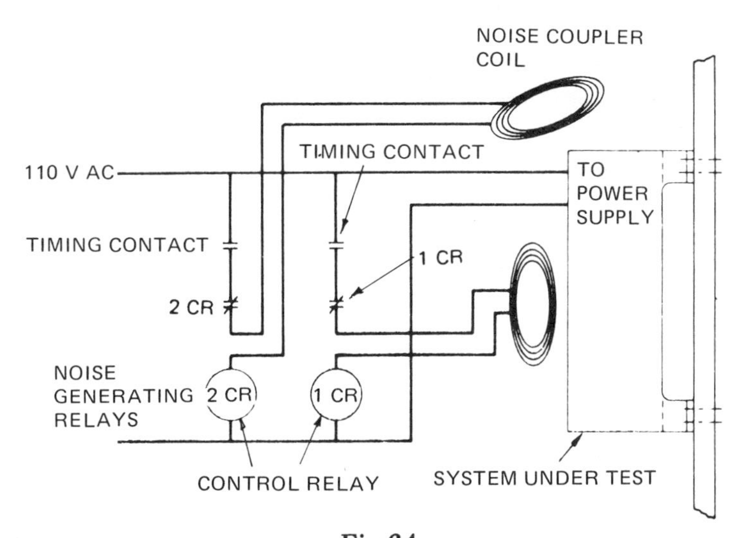

Fig 64
Relay Noise Test Components for Simulating Industrial Noise Environment

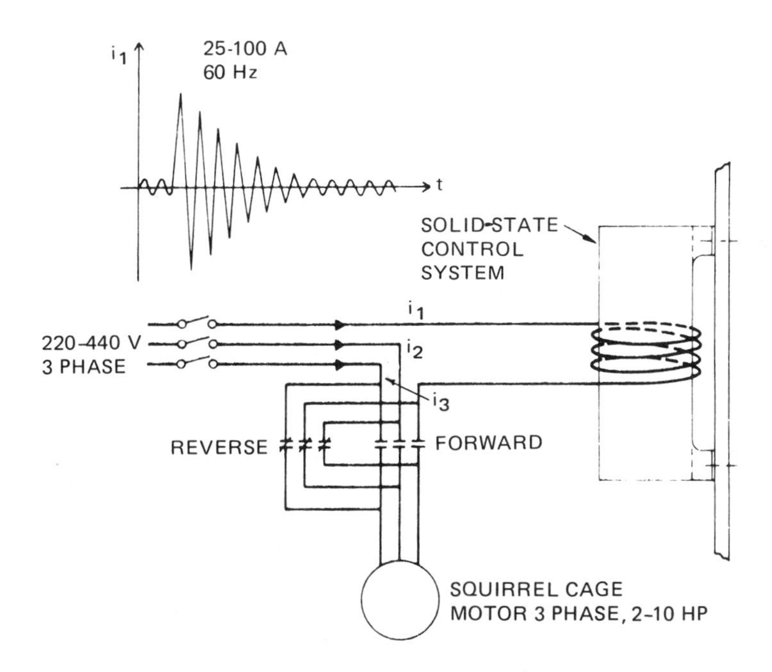

Fig 65
Motor Noise Test Applied to Shielded Control System Simulates Electromagnetic Field by Switching Squirrel-Cage Motors from Full Forward to Full Reverse

Fig 66
Typical Motor Test of Industrial Control with Magnetized Windings in Vertical and Horizontal Directions

speed of the integrated-circuit chips increases. The same is true for the leakage of the semiconductors. Both phenomena result in a reduction of noise immunity.

With the shield design method one first determines the noise sensitivity of the system without shields. Then ferromagnetic shields are provided, and the tests are repeated to establish the degree of shielding effectiveness.[4]

5.3.3 Conducted Noise Test or Line-Propagated Noise Test. The test is required to evaluate the noise immunity of power supplies with regard to filter effectiveness and direct noise coupling. The generated noise spike is superimposed on the 60 Hz power feed line and is represented by a pulse with a rise time of less than 1 μs and a damped oscillation with a fall time of approximately 50 μs. The test apparatus permits variations of the magnitude and different phase synchronisms with the line frequency (60 Hz) and therefore allows for the first time some quantified evaluation of line-propagated noise interference.

[4] Landis Tool Company, Waynesboro, PA, brought this type of test to our attention.

5.4 References

[1] ANSI/IEEE Std 100-1977, Dictionary of Electrical and Electronics Terms.

[2] SCHLICKE, H.M. *Electromagnetic Compossibility*. (Applied Principles of Cost-Effective Control of Electromagnetic Interference and Hazards), 2nd Enlarged Ed. New York: Marcel Dekker, 1982.

6. Installation Recommendations and Wiring Practices

6.1 Basic Elements of Electrical Noise. This section is intended to stand by itself. Furthermore, it is written for the installer of equipment, not the designer, and as such it will deal with the subject in a subjective manner. Reference is made to previous sections for supporting technical background.

For purposes of this guide electrical noise will be defined as an unwanted electrical signal, which produces undesirable effects in the circuits of the control systems in which it occurs. As such it may cause serious malfunctioning of exposed electrical equipment.

In its simplest terms, before an electrical noise problem can exist, there must be three basic elements:

(1) A source of noise
(2) A means of coupling
(3) A circuit sensitive to the noise

The simplicity in this three-element problem is that the problem can be eliminated by removing any one of the three elements. By that it is meant that if an installation is malfunctioning because of electrical noise, it can be made to work correctly by suppressing the noise at its source, *or* by eliminating the way the noise is getting into the system, *or* by desensitizing the equipment and making it insensitive to the noise.

These alternatives are most significant when it is recognized that the speed and costs of the solution will vary with each installation. At one extreme electrical noise is truly an area shrouded in confusion, misunderstanding, and perhaps a little "black magic." At the other extreme it is a phenomenon that is relatively easy to live with if some rather straightforward precautions are taken.

The first step in eliminating electrical noise problems is to recognize that they exist. Technological advances are fundamentally aimed at making equipment smaller, faster, or more sensitive, and this effort makes the concern for electrical noise even greater.

It is impossible to describe a practical guide for the installation of equipment which will be perfect 100% of the times it is used. However, most of the installations will be made in a trouble-free fashion, even if only the more fundamental concerns are considered.

As a starting point it must be presupposed that there are no internal noise problems with the equipment before external wires are connected to it. It is therefore the connections and interconnections where it is believed that special steps need be taken.

Likewise, since the supplier has little or no control over the electrical, mechanical, and geographical limitations of the installation, the user must carefully adhere to the manufacturer's recommendations. In other words, both the user and the manufacturer have responsibilities to themselves and to each other.

6.1.1 Sources of Electrical Noise. In its simplest form one can say that any electrical signal is a potential source of noise to any other signal. There are obviously noise sources that are to be avoided more than others, but this is so completely interrelated with sensitivity that it is very difficult to make a recommendation covering all situations.

In general the higher the voltage or current or the faster the rate of change of either, the more these sources of noise should be avoided by wires which are susceptible to noise.

The following noise sources are most likely to result in electrical noise problems (see 4.1):

(1) Inductive devices, particularly relays, contactors, and solenoids

(2) Ac and dc power lines

(3) Switchgear

(4) Fast rise time sources, principally thyristor-type power devices and, to a lesser extent, solid-state switching circuits and digital circuitry

(5) Variable frequency-variable current devices.

6.1.2 Means of Coupling Electrical Noise. In the overall attempt to minimize the noise or unwanted electrical signals, an effort to minimize the coupling is the most practical technique to eliminate electrical noise problems.

The following lists effective methods of reducing electrical noise coupling:

(1) Maintain physical separation between electrical noise sources and sensitive equipment and between electrical-noise-bearing wires and sensitive signal wires. This is the most effective electrical noise reduction technique.

(2) Twisted pair wiring should be used in critical signal circuits and in noise-producing circuits to minimize magnetic interference (see 6.4).

(3) Shielding practices should be followed as outlined in 6.3.

(4) Proper grounding practices should be followed.

(5) In case of conductive coupling, use appropriate filters (see 6.5).

6.1.3 Sensitivity to Noise. In many ways the susceptibility to noise can be considered in terms identical to "source of noise," namely, the lower the normally expected level of voltage or current, the more likely it is that these circuits are susceptible to noise.

Prior to 1960 industrial control systems were seldom of the fast-responding high-performance variety. As such, electrical noise interference was generally not a problem.

Since 1960 advancing technology dictates fast-response, high-performance, and trouble-free systems. This type of requirement simply cannot tolerate electrical noise of any significant magnitude.

Users and installers of electrical equipment can contribute in several ways to minimize the effect of electrical noise.

(1) Follow the signal-wire instructions supplied with the equipment.

(2) Where signal-wire instructions are not supplied, request them.

(3) When equipment must be installed in an electrically noisy environ-

ment, notify equipment suppliers in advance. Most, if not all, of the problems can be worked out prior to the installation.

(4) If the user prefers specific types of conduit, wires, trays, etc, the equipment supplier should be notified in advance.

6.2 Grounding and Bonding

6.2.1 General

6.2.1.1 Grounding Coordination. The grounding of control equipment in an industrial environment should be integrated by the equipment user into a coordinated ground system. System grounding refers to grounding associated with the power distribution system. When control equipment is grounded, it is interconnected into the system ground by building columns, ground rods, etc. Insulation failures can result in fault current or in the imposition of a ground into the control equipment at some point. For example, if the equipment is powered by a three-phase three-wire system, such as from a transformer secondary connected in a delta configuration, then a phase-to-enclosure fault, assuming that the enclosure is grounded, will result in a ground being imposed upon the ac input system. No fault currents will occur unless a second phase-to-enclosure fault occurs, although arcing is possible at the first fault area due to capacitance charging currents. In either case, fault current levels or ground fault detection schemes, or both, should be considered when equipment is grounded. Personnel protection and maintenance procedures are also factors to consider in association with equipment grounding. For a more detailed discussion of power system grounding see [2].[5]

6.2.1.2 Grounds in a Control System. In a control system there may be a multiplicity of subsystems tied into a ground system. Initially the noncurrent-carrying metallic enclosure shall be grounded; this is an equipment ground. The input ac power system may or may not have a system ground. The input ac may be used to create other ac power systems which again in turn may or may not be tied into the ground system. In addition, the electrical and electronic hardware which provides control over the power flow generally has a zero potential reference point to which the hardware in a localized area is referenced; this point is called a control common.

Figure 67 shows some of the power distribution that may exist in control equipment. Primary power flow is shown going through some type of control element. There are many paths for auxiliary power systems which can be grounded or ungrounded or referenced to the control common.

Figure 68 shows a typical control system in which the electronic part of the control equipment is completely isolated from the main power distribution system by transformers. When a cabinet enclosure is grounded, it is tied into a ground system. This grounding may tie into an input distribution system ground, or other separately derived systems may be tied to ground by this connection. In addition, it is necessary that the control common of the

[5] The numbers in brackets correspond to those of the references listed in 6.7.

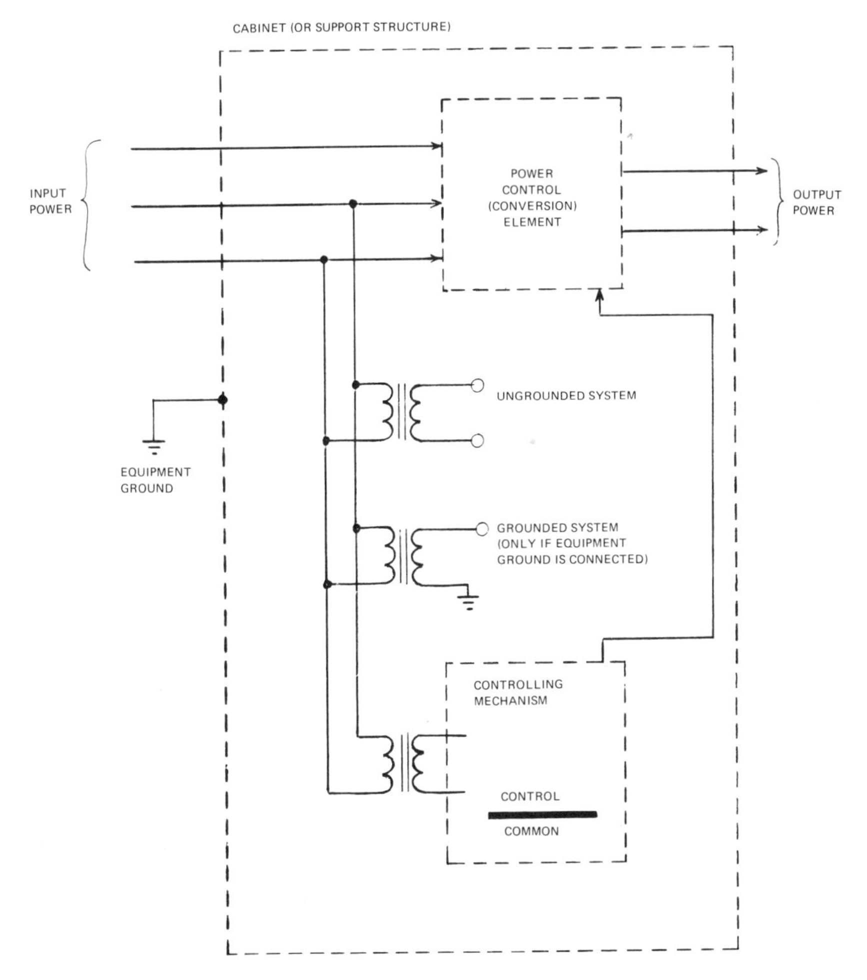

Fig 67
Some Power Distribution which May Exist in Control Equipment

electronic equipment be tied into the ground system in a predetermined fashion.

6.2.1.3 Grounding Philosophy. The design should be based on the concept of two separate electrically insulated ground systems, as discussed in 6.2.1.2. These two ground systems are the equipment ground system and the control common system. The manner in which the two systems are grounded will depend upon the nature of the equipment. The objectives sought by these grounds are as follows:

(1) To minimize the effects of electrical noise which exists in an industrial environment.

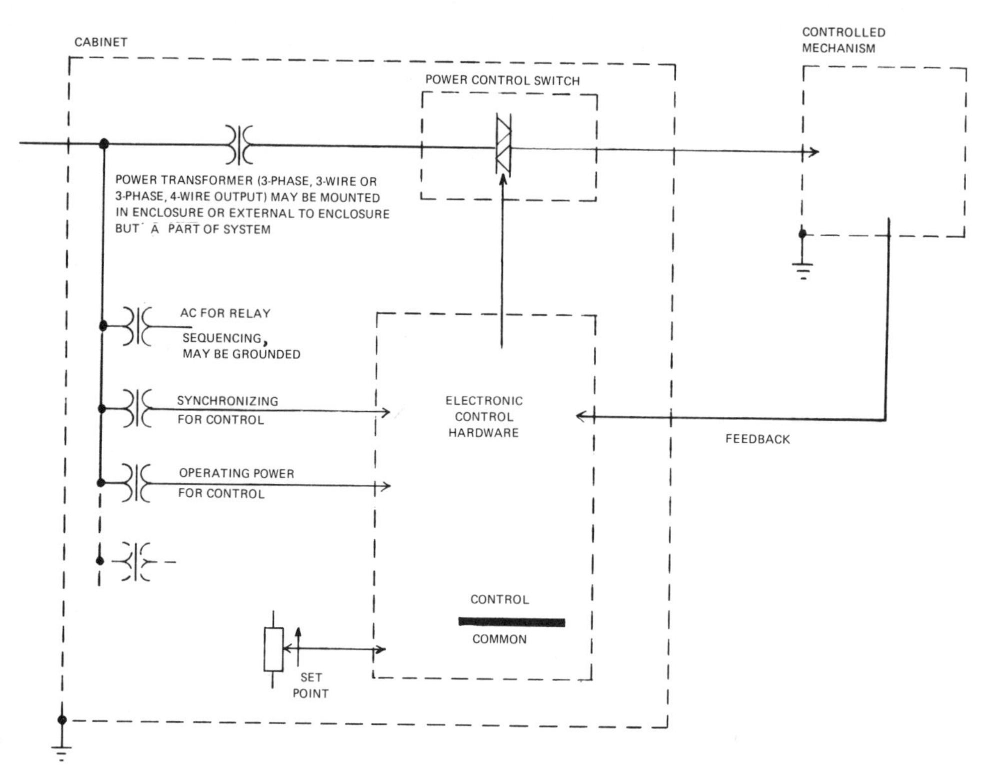

Fig 68
Typical Control System in which Electronic Part of Control Equipment is Completely Isolated from Main Power Distribution System by Transformers

(2) To minimize any disturbing effects within the control equipment and any propagation effects to physically associated equipment which might result from ground currents or excessive fault currents that might occur.

(3) To minimize the shock exposure potential which might appear on non-current-carrying equipment in the event of an insulation failure which might attempt to elevate the enclosing structure to a dangerous voltage level. The resultant cabinet voltage which may appear due to an insulation failure of either an ac or a dc power line can be minimized by proper grounding of the equipment.

Within any physical structure there should be two designated ground reference points, the equipment ground reference point and the control common reference point. The purposes of these two points are first to ensure separation of the ground systems within the structure, and second to provide single connection points to which external connections to the plant grounding system are made.

By separating the ground systems within a cabinet enclosure, ground loops within the structure can be eliminated. On exiting the enclosure, there are different methods which might be chosen depending upon the nature of the equipment. Some of the reasons for the different grounding methods are explained in the following examples.

Figure 69 shows a typical control system for a motor in which the control common is tied to the cabinet at a selected point, which in turn is tied into a ground system via the building column. By separating the two systems within the enclosure, fault currents which might occur on the power system are prevented from traveling on wires that are not designed for such duty. Ground loops within the equipment are also eliminated. If the system is somewhat limited from a physical size standpoint, the potential of the enclosure with respect to a ground plane is not important from an electrical standpoint, that is, relative voltage changes between the control common/ cabinet enclosure and the external ground plane from a potential standpoint should not affect the operation of the control equipment. From a safety standpoint, the potential differences which might occur must be considered.

Figure 70 shows the same control equipment as Fig 69, except that the control common and the equipment ground points are separated to point out possible noise sources. As the cabinet enclosure is electrically coupled to the control equipment through distributed capacitance, ground currents can create an external signal which changes the relative potential between the cabinet enclosure and the control common. Signals of this nature might be coupled into the equipment.

Figure 71 shows a typical control system for three motors which might perform a process-line function. In this case each of the control commons is tied at a preselected point to the cabinet enclosure, which in turn is tied into the plant ground system, for example, a building column. Ground currents in the ground system may cause potential differences to exist between all three cabinet enclosures. In addition, fault current exiting the equipment via the equipment ground will change the reference level of the faulted equipment. Depending upon the nature of communications between the cabinets, potential differences due to ground currents or fault currents may create serious system problems.

Figure 72 shows the same control system for three motors, except for the method of grounding. Each cabinet is tied to the ground system by the building columns. The control commons are tied to the ground system by a grounding electrode which provides a zero reference point for the control common in each of the three cabinets. Potential differences between the enclosure and the control common due to high-frequency high-energy transients created by power system ground currents should be minimized by the shunt capacitor. The capacitor selected must have good high-frequency characteristics in order to minimize the effects upon digital and linear circuitry of the externally generated disturbances in the ground system. Extended foil capacitors can be used in this type of application. Typical values are 0.47 μF.

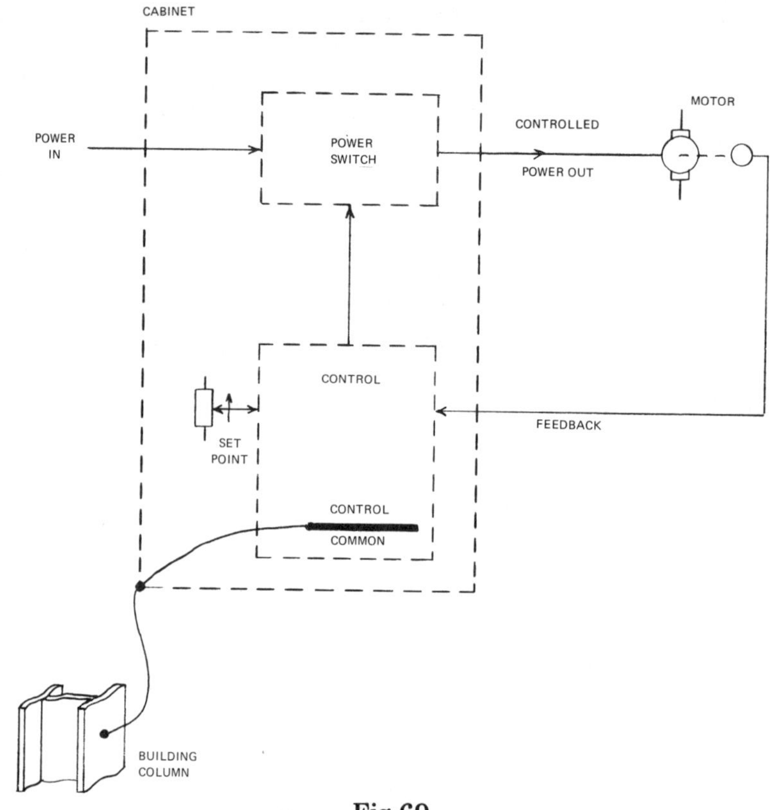

Fig 69
Typical Control System in which Control Common Is Tied to Cabinet at Selected Point which in Turn Is Tied into Ground System via Building Column

The voltage rating must exceed the voltage that can be created by fault currents within the structure.

The equipment ground should connect the mechanical cabinet into the plant grounding system. The connection is made from the designated equipment ground reference point, which should be tied to the building ground, such as a building column. The equipment ground will prevent hazardous potentials from developing between adjacent equipment enclosures and between enclosures and building ground. The control common reference point is connected internally within the cabinet to the control common(s). The external connection from this reference point will vary depending upon the system. The control common reference point could be tied, for example, to an earth ground which could consist of a copper rod driven into the earth (grounding electrode), a building column embedded in the earth, or any other mass which has been determined to be at *true earth* potential. The point of connection to this earth ground would be designated as the control

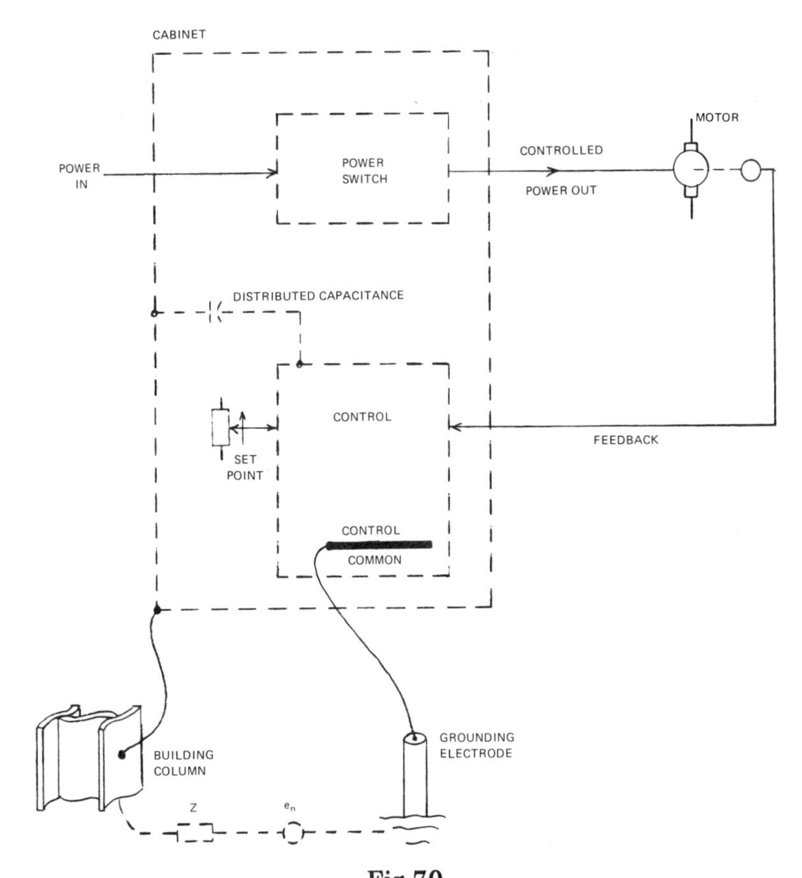

Fig 70
Same Control Equipment as in Fig 69, Except that Control Common and Equipment Ground Points Are Separated to Point out Possible Noise Sources

system zero reference point, which would allow all control system signals to be references to the same point. The distinction between the term *common* and the term *ground* must always be kept in mind. Under normal operating conditions, a common is intended to carry current, while a ground is not. Grounds must be designed to carry fault currents under abnormal conditions. The control common is a ground and should be treated as both a common and a ground. Minimum wire sizes, color coding of wire, general safety practices, etc, should comply with ANSI/NFPA-1981 [1].

6.2.2 Equipment Ground

(1) Each piece of equipment should be constructed such that a grounding connection can easily be made to the equipment enclosure, frame, etc, so as to ensure good, electrical contact. The term *piece of equipment* means a mechanically integrated structure made of non current-carrying material, be it a single cabinet structure or a collection of cabinet structures connected

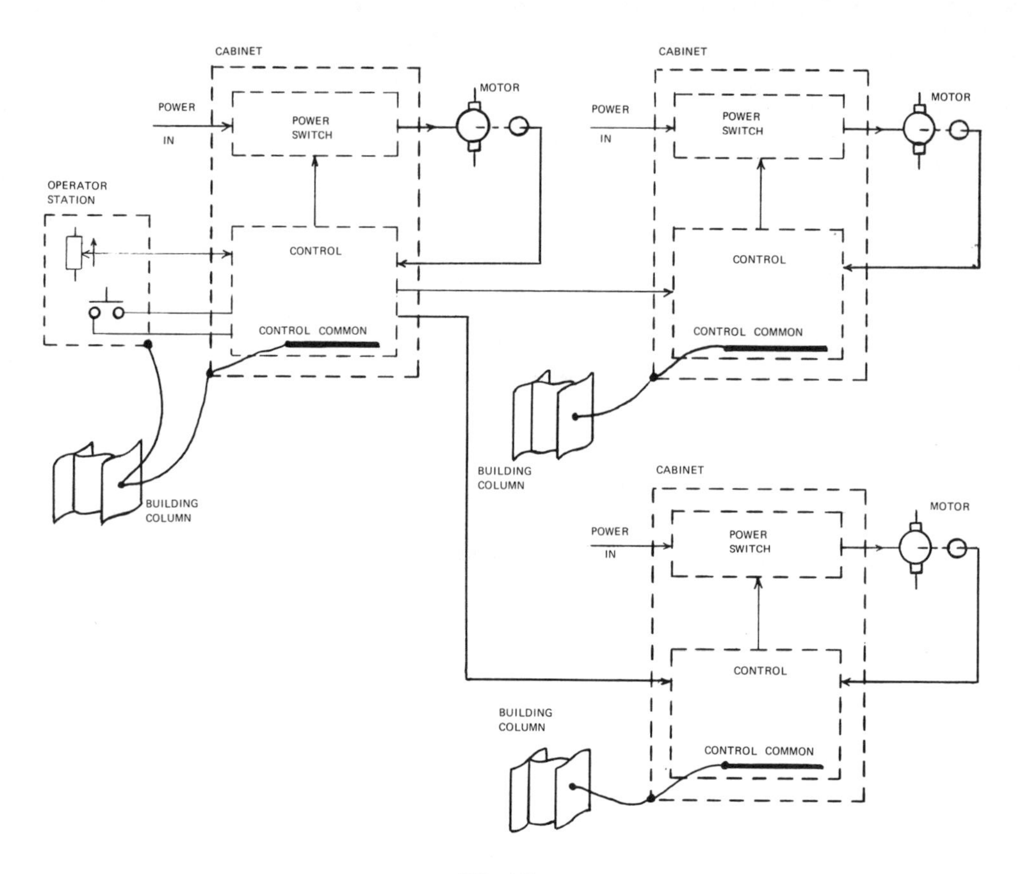

Fig 71
Typical Control System for Three Motors which Might Perform Process-Line Function

together. If the various parts of the mechanical structure have been painted, it is assumed that proper procedures are used in the mechanical integration to ensure positive electrical contact.

(2) Where possible, the only connection from cabinet to ground should be from the designated equipment ground reference point.

(3) All cabinets, enclosures, motor frames, etc, should be tied to the installation grounding system in one of the following ways:

(a) If the cabinet is electrically insulated from ground, then a single connection should be made from the designated equipment ground reference point located in the cabinet structure to the building ground system via an insulated conductor of appropriate size and of stranded or braided construction.

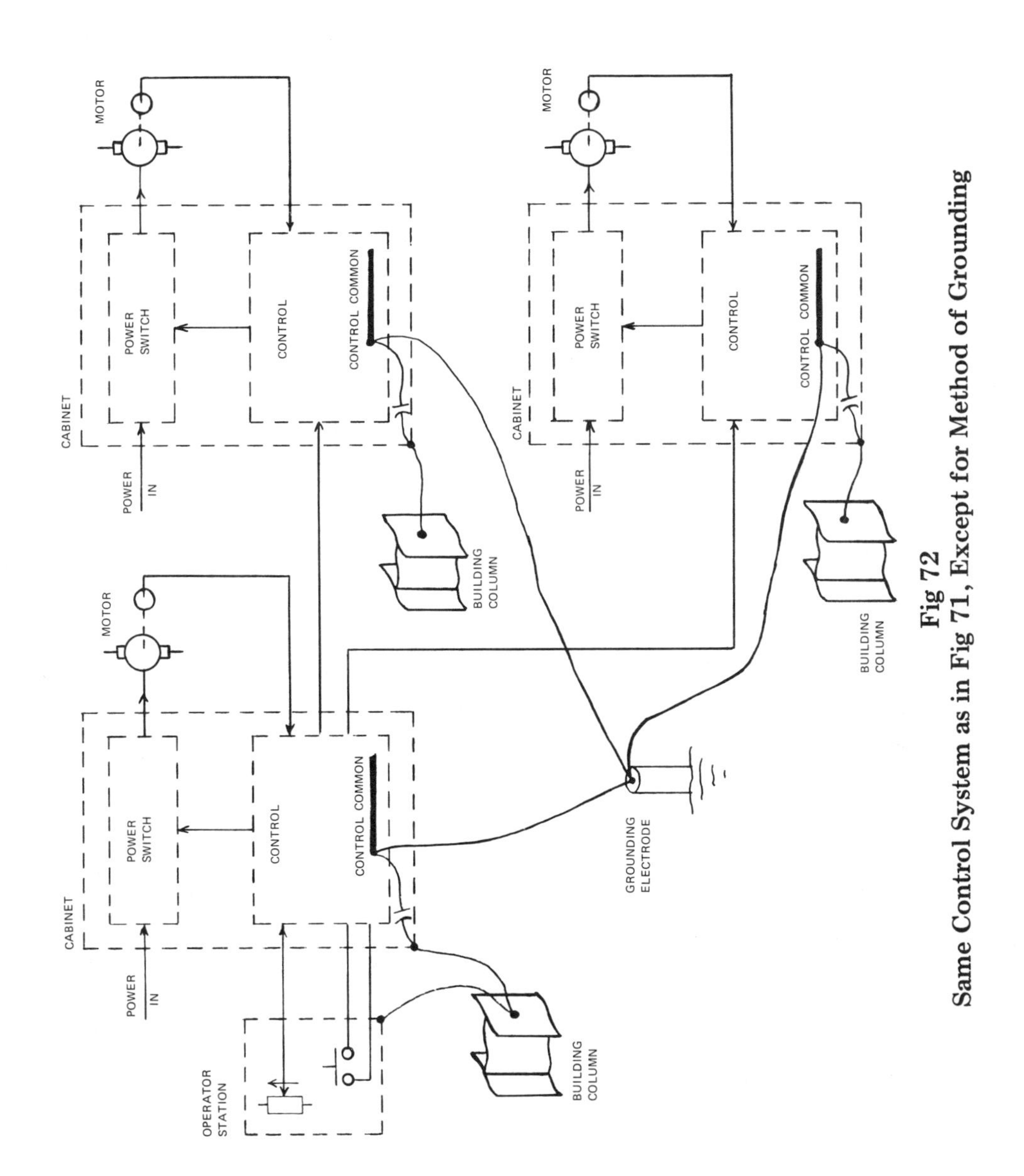

Fig 72
Same Control System as in Fig 71, Except for Method of Grounding

(b) Otherwise the enclosure should be referenced to the nearest building structure. This connection should be made with an insulated, braided strap or a stranded conductor.

(4) All individual chassis (in particular sliding chassis which remain electrically operable regardless of position) which are not mechanically integrated into the structure so as to ensure good electrical contact should be connected to the mechanical structure by braided insulated strap.

(5) In multiple-component systems which are not mechanically integrated, the individual equipment ground reference points should be tied to a single common reference point, such as a centrally located building column.

(6) Underground cold-water pipes should be used as ground electrodes according to ANSI/NFPA 70-1981 [1], Section 250-81.

6.2.3 Control Common (Ground)

(1) Each piece of control equipment should have a control common which is insulated from all chassis (mechanical structure).

(2) Each control common should be electrically connected to a designated control common reference point with appropriately sized cable.

(3) As has been indicated, there are various methods for connecting the control common reference point into the ground system, and these methods are detailed in the National Electrical Code [1], Article 250. Some of these methods are listed below:

(a) Directly to a building column or electrically conducting mass which is determined to be at *true earth* potential. This may or may not be the same point to which the equipment ground reference is connected.

(b) To the termination point of the equipment ground reference point.

(c) To the equipment ground reference point itself.

Grounding of the control common reference point depends upon the equipment size and means of communication between cabinets which, although a part of a control system function, are physically separated. Manufacturers' recommendations with respect to grounding should be followed.

(4) Both the equipment ground and the control common reference points should have no more than one lead exiting the equipment enclosure. This lead should be of standard construction. The minimum size will be determined by the system. Precautions should be taken to ensure a good permanent electrical connection from the ground system to the grounding conductors.

(5) When recommended by the manufacturer, every major system should have its own individual grounding electrode to be used solely for control common ground purposes. This point then becomes the zero-potential reference point for every other point in the system.

(6) The following is pertinent to grounding electrodes:

(a) If grounding electrodes are used it is assumed that they are designed as a part of the total plant grounding system.

(b) It is assumed that the functioning of the ground rods is checked at the time of installation and at prescribed maintenance intervals.

(c) Procedures for measuring earth ground resistance are discussed in [1].

6.2.4 Documentation. The following documentation should be available as part of the system documentation wherever and whenever possible:

(1) A list of the devices whose inputs or outputs, or both, are not tied into control common

(2) A list of the devices whose inputs or outputs, or both, are tied to the equipment grounding conductor

(3) An ac and a dc equipment ground wiring distribution diagram

(4) A control common wiring distribution diagram

(5) A wiring diagram of all direct connections between control common and the power ground or the equipment enclosure

6.2.5 Buses. All control common buses should be sized to keep the *IR* drop at a minimum throughout the system.

6.2.6 Bonding. In a bolted-frame constructed cabinet all mechanical pieces should be electrically interconnected or bonded in a positive manner to ensure personnel safety via the equipment ground connection.

6.2.7 Ground Loops. In any system communication with a physically separated part of the control, careful consideration should be given to interconnections so as to prevent the inadvertent generation of ground loops. Relays, optically coupled devices, magnetic amplifiers, and differential input voltage amplifiers can be used to eliminate ground loops.

6.3 Shielding Practices

6.3.1 Shielding against Electric Fields. An effective electrostatic shield should (1) completely enclose the shielded components, (2) be held at a constant potential, preferably at the reference potential of the system, and (3) have high conductivity.

In practical cases it is sometimes difficult to completely isolate a conductor from the others. In such cases it is a good practice to tie all isolated and unused conductors to ground (or the reference bus). This will increase the effective shielding.

6.3.2 Shielding against Magnetic Fields. An effective magnetic shield should (1) completely enclose the shielded components and (2) have high permeability. Magnetic shielding is sometimes more difficult than electrostatic shielding because it is relatively easy to obtain high conductivity in an electric shield as compared with the ability to obtain a high permeability and a high conductivity in a magnetic material.

6.3.3 Shielding against High-Frequency Fields. There is no perfect shield against high-frequency fields. Upon impinging on the shield surface, some part of the incoming wave is reflected back and the rest travels through the shield, being absorbed (attenuated) by the shield as it travels along.

The absorption losses are proportional to the square root of the frequency, the conductivity, and the permeability of the shielding material. The reflection losses are different for the different types of disturbing field (plane wave, electric wave, and magnetic wave). The reflection losses are also dependent on the frequency, and on the conductivity and the permeability of the shielding material. However, the relationship is more complex for the reflection losses than for the absorption losses. A material of high conductivity is a good reflector, and a material of high permeability is a good absorber.

Multiple shielding can be applied in many cases with considerable advantage. Even if the same shielding material of the same total thickness is used, multiple shielding has higher shielding efficiency than the single shield because of the additional reflective surfaces. However, it may not be practical because of the necessity of air gaps required for such construction. This problem can be avoided by using a shield that is composed of two different metals laminated to each other. It is advantageous to use a shield consisting of composite layers of copper and a ferromagnetic material, with the copper layer toward the source to utilize the substantial reflection losses at the air-copper surface. The presence of the ferromagnetic material would produce higher absorption losses.

6.4 Wiring Practices

6.4.1 Wire Selection

6.4.1.1 Wire Size. The wire selected for supplying power to industrial controllers and for interconnecting between various parts of the control system must be sized to meet the current-carrying requirements (ampacity) for that application. The National Electrical Code [1] contains tables which may be used as a guide. The voltage classification of the wiring should equal or exceed the rated circuitry voltage.

The wire must be able to withstand the physical stresses to which it may be subjected during installation or operation. The size of the termination, for example, the terminal block, may limit the maximum size.

Multiconductor groupings within a single jacket may be used to attain increased strength of small conductors.

6.4.1.2 Wire Type. The type of wiring to be used for interconnecting parts of the control system depends on the noise susceptibility of the components. The system supplier will furnish specific instructions if unusual or highly sophisticated wiring is required. The following general guides may be used if such instructions are not furnished.

6.4.1.2.1 Highly Susceptible Systems. Twisted wires should be used for power feeds. Interconnecting control wires should be twisted pairs, provided that the signals are of dc on relatively low frequency. Higher frequency signals such as 10 kHz and greater should be carried in coaxial-type cables.

6.4.1.2.2 Low-Susceptibility Systems. These systems may be normal industrial practice utilizing either multiconductor cables or signal conductor wires grouped together.

Control circuitry wiring should be physically separated from power wiring.

6.4.2 Wire Shielding Practices. Noise by definition is an unwanted signal. Shielding by definition is the protection of the signal wires from noise, or unwanted signals.

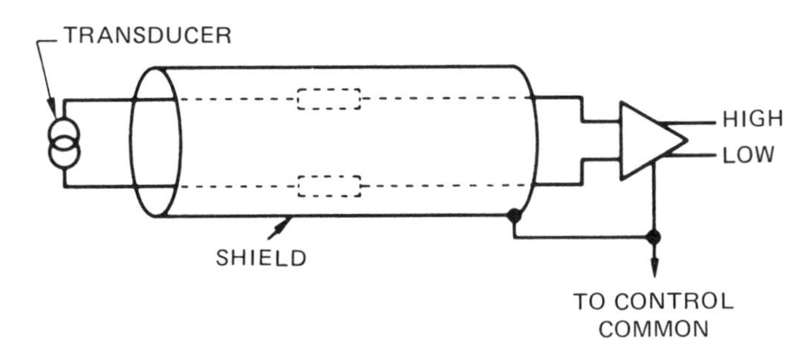

Fig 73
Typical Shield

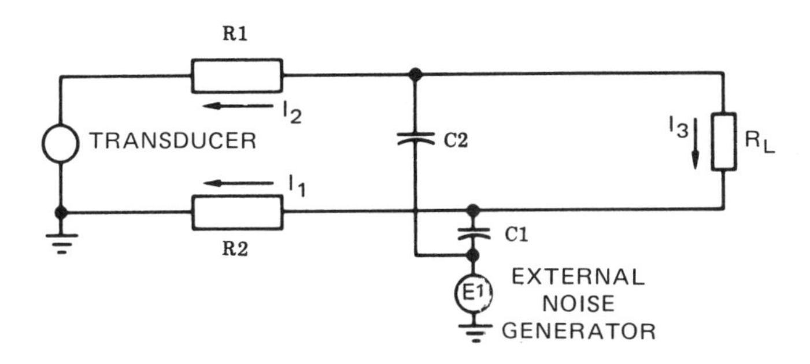

Fig 74
Electrostatic Coupled Noise

Noise is normally introduced into the signal circuits through electrostatic (capacitive) coupling, magnetic (inductive) coupling, and resistance coupling. The reduction of these noise signals takes the form of shielding and twisting of signal leads, proper grounding, and separation.

The purpose of the shield is to reduce the magnitude of the noise coupled into the low-level signal circuits by electrostatic or magnetic coupling. The shield may be considered an envelope surrounding a circuit (see Fig 73) so as to reduce this coupling.

6.4.2.1 Electrostatic Coupling. Electrostatic or capacitive coupling of external noise is shown in Fig 74. The external noise source couples the noise into the signal wires through capacitors C1 and C2, and the resulting flow of current produces error voltage signals across R1, R2, and R_L. The error signal is proportional to the length of the leads, the impedance of the leads, the amplitude and frequency of the noise signal, and the relative distance of the leads from the noise source.

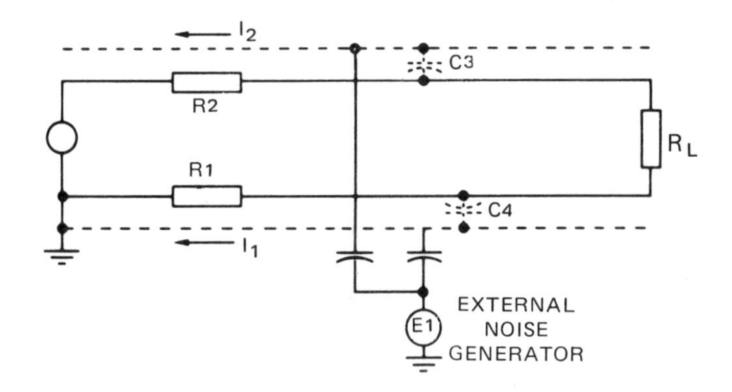

Fig 75
Use of Shield to Reduce Electrostatic Noise

The noise due to electrostatic coupling can be reduced by the use of shielded wire, by separation, and by twisting of leads. As the separation between the noise source and the signal wires is increased, the noise coupling is thereby reduced. Twisting of the leads provides a balanced capacitive coupling (which tends to make C1 = C2), reducing the noise level.

The use of a shield to reduce electrostatic noise is illustrated in Fig 75. The noise-induced currents now flow through the shield and return to ground instead of flowing through the signal wires. With the shield and signal wire tied to ground at one end, a zero-potential difference would exist between the wires and the shield. Hence no signal current flows between wire and shield.

6.4.2.2 Magnetic Coupling. Magnetic coupling is the electrical property that exists between two or more conductors, such that when there is a current change in one, there will be a resultant induced voltage in the other conductor. Figure 76 shows a disturbing wire (noise source) magnetically coupling a voltage into the signal circuit.

The alternating magnetic flux from the disturbing wire induces a voltage in the signal loop which is proportional to the frequency of the disturbing current, the magnitude of the disturbing current, and the area enclosed by the signal loop, and is inversely proportional to the square of the distance from the disturbing wire to the signal circuit.

Figure 76 shows all of the factors necessary to introduce an error voltage: rate of change of current, a signal loop with a given area, and a separation of the conductors from the disturbing signal.

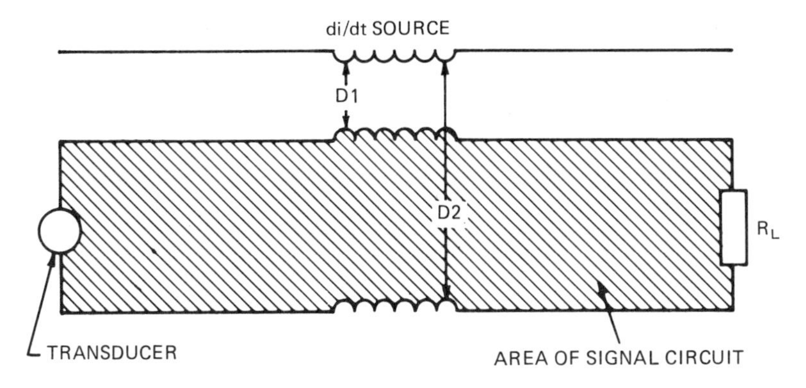

Fig 76
Magnetic Noise Coupling

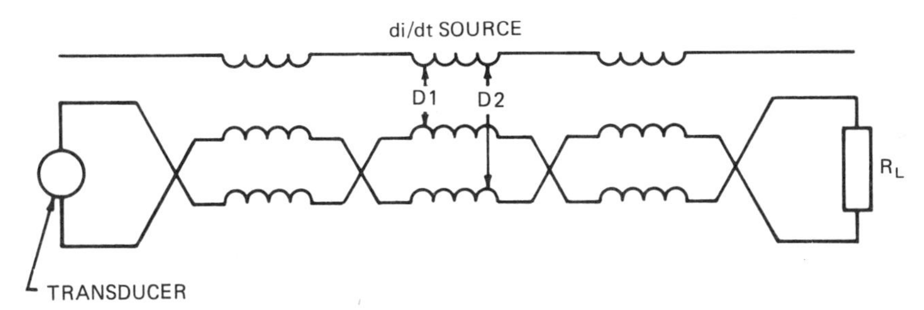

Fig 77
Reducing Magnetic Noise by Twisting of Wires

A common method of reducing the effect of magnetic coupling is the use of twisted conductors in the signal circuits, as shown in Fig 77. The distance of these two signal wires with respect to the disturbing wire is approximately equal, and the area of the circuit loop is almost zero. Reducing this area to practically zero will reduce the voltage induced by the magnetic field to almost zero due to the equal magnitude of current induced in each lead resulting in a near zero net circulating current.

Table 4 [3], [6] will give the installer the relative effectiveness of employing twisted wire to reduce magnetic interference.

Magnetic coupling can also be reduced by employing a shield around the signal wires. The shield is effective because the magnetic field produces eddy currents in the shield which oppose the original magnetic field. A sketch of this type of noise reduction is shown in Fig 78. The shield can be either magnetic or nonmagnetic, depending upon the type of noise rejection desired. For truly effective shielding against magnetic fields a special magnetic material must be used.

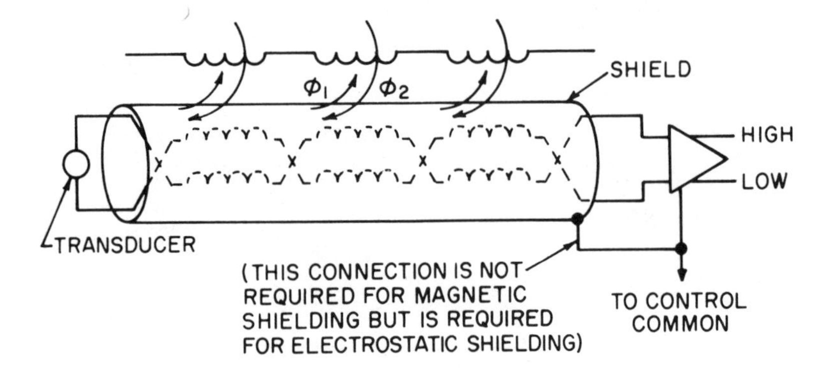

Fig 78
Effects of Shield in Reducing Magnetic Coupling

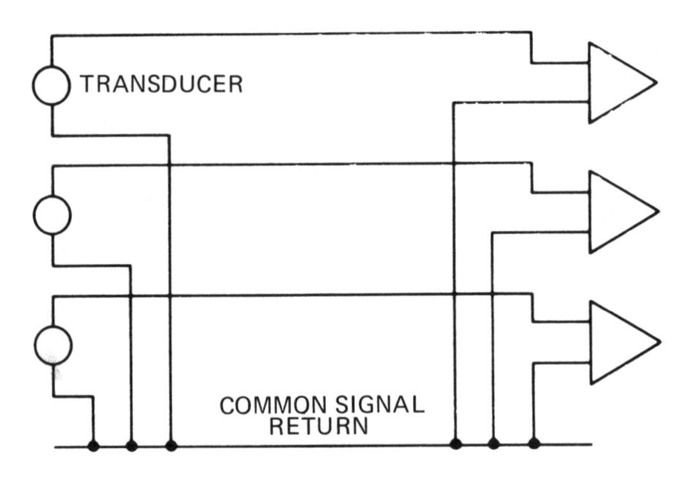

Fig 79
Resistance Coupling

6.4.2.3 Impedance Coupling. Impedance coupling is the electrical property that exists when two or more signal wires share the same common return signal wire such that when there is a signal current in one of the signal wires, there will be a resultant signal voltage in the others.

To avoid impedance coupling in signal circuits one can do either of the following:

(1) Employ low-resistance wire or bus for the common return when a common return cannot be avoided. (For critical applications both the resistance and the inductance of the bus should be minimized.)

(2) Whenever possible, employ separate signal return leads.

Figure 79 shows an example of impedance coupling. The common impedance return, if necessary, should be low impedance with respect to the signals being carried by the return wire or bus.

6.4.2.4 Classification of Shields. In considering shielding, there are four distinct classifications, the shielded wire or cable, the wire carrier or cable carrier, the equipment shield or guard, and the component shield.

6.4.2.4.1 Shielded Wire. It consists of a twisted wire pair or pairs protected by a shield. To be effective, a continuous foil or metalized plastic shield is preferable. The shield should be insulated and equipped with a copper drain wire for convenient single-point connection.

6.4.2.4.2 Wire Carrier. It takes the form of a conduit or cable tray to protect and isolate the wiring.

6.4.2.4.3 Equipment Shield. It takes the form of a cover or guard employed to isolate sensitive regulator circuitry from external sources or noise.

6.4.2.4.4 Component Shield. It takes the form of a shield on individual components such as transformers, which are sometimes equipped with internal shields to provide low-impedance paths for electrostatic noise signals.

6.4.2.5 Shielding Material

6.4.2.5.1 Shielded Wire. Table 4 illustrates the effectiveness of using twisted pair wires in reducing magnetic field noise in signal circuits. Good noise cancellation is provided by 1 inch lay or 12 twists per foot. Theoretically the use of twisted wires could nearly eliminate both electric field and magnetic field noises. Some manufacturers have gone to 0.5 inch lay to obtain truly balanced twisting so as to avoid the use of shielded wire.

Table 5 [3], [6] shows the effectiveness of various types of shielded wire in reducing electrostatic field noise in signal circuits. As can be seen from the table, lapped or taped shields provide more shield coverage and are more effective. Because these shielded wires are relatively ineffective against magnetic field interference, twisted wire should be employed within the shield. Because of manufacturing expense, only 1.5 inch and 2 inch lay wires are available within the shield. This combination has been found to be more than adequate for most applications. An insulated jacket should be placed over the shield so that the shield may be grounded at one end only.

Coaxial cable is used extensively for critical circuits since the voltages induced in it by external magnetic fields essentially cancel out, and the outer

Table 4
Magnetic Interference Reduction

Type	Noise Reduction (ratio)	(dB)
Parallel wires	—	0
Twisted wires		
4 in lay	14:1	23
3 in lay	71:1	37
2 in lay	112:1	41
1 in lay	141:1	43
Parallel wires in 1 in rigid steel conduit	22:1	27

Table 5
Electrostatic Noise Test Results

Shield	Noise Reduction (ratio)	(dB)
Copper braid (85% coverage)	103:1	40.3
Spiral wrapped copper tape	376:1	51.5
Aluminum - mylar tape with drain wire (total coverage)	6610:1	76.4

No shield = 0 dB.

conductor shields the inner conductor from electric fields, all as a result of the concentric construction of the cable. The only shortcoming of this cable is the rather large capacitance to ground which could result in an ac ground loop. This undesirable capacitance can be reduced to a negligible value by using triaxial cable which has a second, or overall, shield. Both of these types of cable are considerably more expensive than twisted wires or shielded twisted pair cable. This cost, plus the greatly increased raceway area required for a given number of circuits, usually limits the use of this cable.

Figure 80 shows a double-shielded wire which is employed on very-low-level signal circuits. The inner signal wires are twisted pairs. The inner shield is usually grounded at one point, whereas the outer shield may be grounded at more than one point to eliminate radio-frequency circulating loops.

6.4.2.5.2 Raceway Shielding. In the industrial environment the conduit or raceway is employed as a carrier of signal and power wiring, which provides the communication link between the process and the controller. This communication link may be short or run for several miles through an interference environment. The choice of how the communication link is run and what materials are employed will greatly influence the degree of success of the installation. Expensive downtime can be avoided by properly designing the communication link.

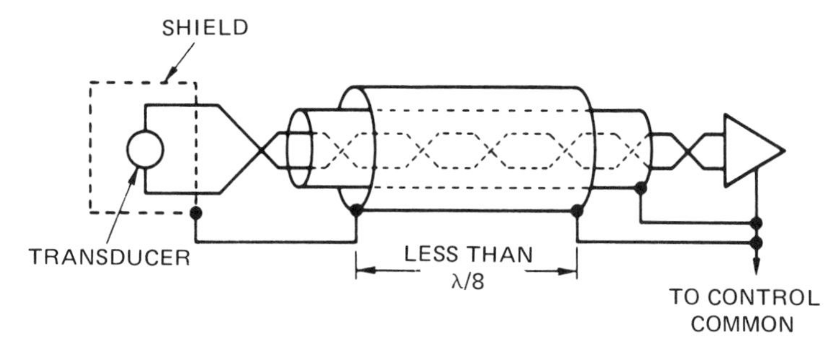

Fig 80
Double-Shielded Conductor

Table 6
Raceway Shielding

Raceway Type	Thickness (in)	60 Hz Magnetic Field Attenuation (ratio)	(dB)	100 kHz Electric Field Attenuation (ratio)	(dB)
Free air		1:1	0	1:1	0
2 in aluminum conduit	0.154	1.5:1	3.3	2150:1	66.5
No 16 gauge aluminum tray	0.060	1.6:1	4.1	15 500:1	83.9
No 16 gauge steel tray	0.060	3:1	9.4	20 000:1	86.0
No 16 gauge galvanized ingot iron tray	0.060	3.2:1	10.0	22 000:1	86.8
2 in IPS copper pipe	0.156	3.3:1	10.2	10 750:1	80.6
No 16 gauge aluminum tray	0.060	4.2:1	11.5	29 000:1	89.3
No 14 gauge galvanized steel tray	0.075	6:1	15.5	23 750:1	87.5
2 in electric metallic tubing	0.065	6.7:1	16.5	3350:1	70.5
2 in rigid galvanized conduit	0.154	40:1	32.0	8850:1	78.9

Table 6 [3], [5], [6] illustrates the relative merits of several types of raceways. The tests for magnetic field attenuation were made at 60 Hz. Tests at higher frequencies indicate that aluminum material usually provides more attenuation than steel at frequencies above 2 kHz.

Table 6 also shows the relative merits of several types of raceways for electrostatic attenuation at 100 kHz. As can be noted, both aluminum and steel provide satisfactory electrostatic field attenuation.

Tables 4 and 6 emphasize the importance of physical separation and twisting of wires to avoid magnetic field interference. Since magnetic field strength is inversely proportional to the square of the distance from the source, doubling the separation could reduce interference up to four times, tripling the separation up to nine times, and so on. Separation also helps to reduce coupling from electrostatic fields.

6.4.2.6 Shielding Usage

6.4.2.6.1 Elimination of Ground Loops. When using a single-shield envelope, the shield should be grounded at only one point. Figure 81 shows that circulating currents can be introduced into the signal wires by having more than one ground.

To prevent ground loops on a single-envelope shield, the following rules should be followed:

(1) The shield envelope should be grounded at only one point.

(2) The shield envelope should have an insulated jacket so as to prevent multiple grounds.

Ground loops can also be prevented by decoupling the input amplifier, as in the case of digital circuits, by the use of an input transformer, as shown in Fig 82. Another technique is to employ optoelectronic couplers to isolate the logic circuits, as shown in Fig 83.

6.4.2.6.2 Shield Connections. The following rules apply when connecting the shields:

(1) The shield should never be left floating since this only enhances the electrostatic field coupling.

(2) As a general rule, low-level signal sources require differential input amplifiers with the location of the shield ground connection to be specified by the equipment supplier.

(3) With double-shielded envelopes, which are employed in critical low-level circuits, it may be necessary to ground the outside conductor at more than one point. The distance between grounding points should be less than $\frac{1}{8}$ wavelength of the expected radio-frequency interfering electrical noise.

(4) Shields of digital logic circuits should only be grounded at the logic power supply.

(5) Multipair cables used with thermocouples must have individually insulated shields so that each shield may be maintained at the particular thermocouple ground potential.

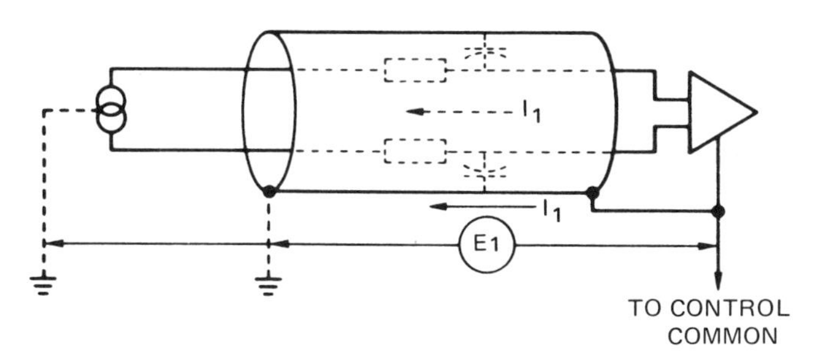

Fig 81
Grounding More than One Point

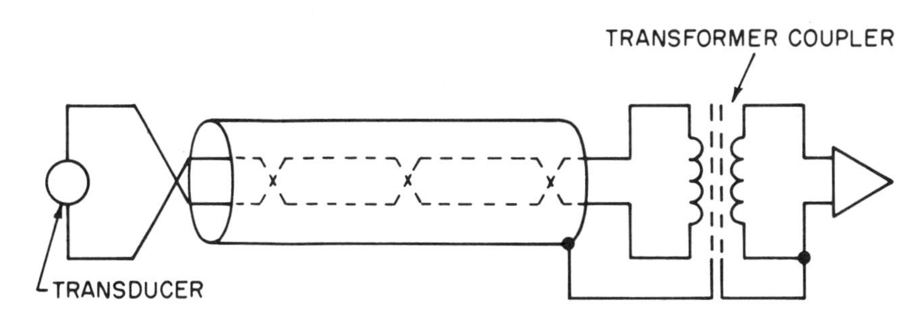

Fig 82
Transformer-Coupled Input

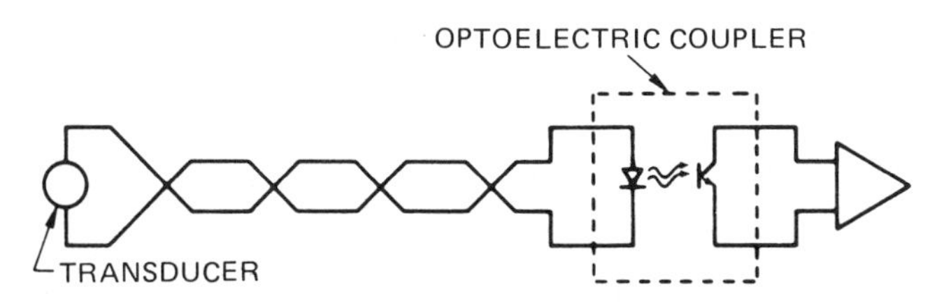

Fig 83
Optoelectronic Coupler Circuit

6.4.2.7 Practical Noise Immunization Techniques and Interconnection Wiring Practices.

(1) In twisted pair signal leads, twisting of the signal *high* lead with its *low* (common, return) lead results in less loop area for inductive pickup and lower inductance per foot. Lower inductance allows higher frequencies to be transmitted over longer distances before signal loss and distortion become significant. In addition, inductive loading effects on the signal source are reduced. The minimal spacing of the high and low (return) signal leads affects cancellation of the magnetic fields of the conductor currents. Cancellation of the conductor fields reduces inductive coupling with other signal leads. For these reasons twisted signal leads should be used wherever and whenever possible in lieu of nontwisted signal leads.

(2) Multiple conductor cable used for the transmission of digital information, such as binary coded decimal (BCD) data, should consist of twisted (but not necessarily twisted pair) conductors.

(3) Multiple-conductor cable used for the transmission of several individual signals should consist of twisted pair signal leads. Each twisted pair should have an electrostatic shield, which is electrically insulated from all other electrostatic shields along the length of the cable.

(4) Multiple-conductor cables should have an overall electrostatic shield.

(5) Unused conductors and electrostatic shields should be terminated. Where multiple-conductor cables are involved, half of the unused conductors and shields should be terminated at one end of the cable and the remainder at the other end of the cable.

(6) Signals transmitted via multiple-conductor cables should have similar characteristics.

(7) For the transmission of signals which must be kept noise-free, cable of special construction is commercially available. This special construction involves specially twisted, specially shielded, or interwoven conductors, or any combination of all three. Three-conductor twisted shielded cable is available for the transmission of three-wire signals such as potentiometer signals (high, low, and wiper) and dc power (for example, +15 V, −15 V, and common) to electronic amplifiers.

(8) Power leads (for example, 120 V ac, 60 Hz equipment power) should be kept closely spaced to maximize the cancellation of conductor magnetic fields.

(9) Where possible, conductors carrying alternating currents should be twisted with their returns.

(10) All electrostatic shields should be terminated. An electrostatic shield should be terminated at one point only. The shield termination point should be at the same electrical potential as that to which the signal is referenced (typically electronic amplifier power supply common).

(11) If a significant potential difference exists between a shield termination point and equipment safety ground, special electrostatic shielding techniques may be required.

(12) A potential difference between electrostatic shields which may make physical contact should be considered undesirable.

(13) Electrostatic shields should be insulated from ground and from each other along their lengths.

(14) The shield (or braid) of a coaxial cable is often used as a signal return. When so used, the shield should be regarded as one conductor of a two-conductor signal lead, and no electrostatic shielding should be assumed. Triaxial cable (electrostatically shielded coaxial cable) is commercially available.

(15) Attention should be given to high-impedance devices [for example, metal oxide semiconductor field effect transistors (MOSFET)] which are susceptible to static charges. Mechanical flexing of the dielectric in a coaxial cable, for example, can produce transient charge effects (analogous to mechanically stressing a piezoelectric crystal).

(16) Where twisted shielded leads must be broken (for example, input-output terminal boards), minimize the untwisted unshielded lead length.

(17) Whenever possible, low-level signal lines should be run unbroken from signal source to signal receiver.

(18) If a shielded signal lead is broken (for example, at a junction box), maintain shield continuity.

(19) Low-level signal leads should not be run parallel with high-current power lines or high-voltage lines.

(20) If signal leads must be run in the vicinity of switchgear, special precautions (for example, use of a localized magnetic barrier) should be considered.

(21) Conduit should not be buried beneath high-voltage power transmission lines or in known ground currents.

(22) Solid-state switches should be given the same noise immunization considerations as digital logic devices.

(23) Particular attention should be given to those signals that will be used in generating difference signals, such as an error-correction signal in a closed-loop control system.

(24) Where signals in the microvolt or microampere range must be transmitted, electrothermal or electrochemical effects, or both, should be considered (for example, terminate copper to copper, minimize thermal gradients along the lead length).

(25) Characteristics of data cables (for example, capacitance, inductance, and resistance per foot) should be documented. Lead capacitance can cause signal degradation and can result in undesirable capacitive loading on a signal source.

(26) All interconnection cable should be documented as to manufacturer, catalog number, capacitance per foot, and so on.

(27) Critical signals should be defined not only in terms of low level or analog, but also in terms of those signals that are absolutely essential to acceptable (as opposed to perfect) equipment operation.

(28) Cable routing should be utilized. Cables should be routed around

rather than through high-noise areas. If two cables carrying signals of different classifications must cross, they should cross abruptly rather than gradually over a long distance.

(20) Data and signal cables should be classified relative to nearby or adjacent cables. If two cables carry data of grossly different currents, voltages, or frequencies, they should be treated as being of unlike classification. Cables of unlike classification should be kept physically separated.

(30) If signals of a like classification are run adjacent to each other, twisted wiring should be used to minimize the possibility of crosstalk between them.

(31) Documentation for a piece of equipment or system should include a list of connections between building column, safety ground, electronic ground, and signal commons, a list of devices whose inputs or outputs are not of a differential or balanced nature, and a list of critical signals or those requiring special noise-immunization considerations. Documentation should be kept up to date.

6.4.3 Cable Spacing. A reasonably strong case has been presented in previous sections of this document for the separation from signal cables of the sources of time-varying voltages dv/dt and time-varying currents di/dt to minimize electric and magnetic coupling. On large control installations the complexity of the communication-system interconnecting cabling presents an enormous challenge if it is to be properly engineered. In order to achieve a separation of electrical noise sources from signal cabling, a classification system has been developed according to equipment-noise susceptibility and noise-generation power level, which will permit an orderly grouping of the cables for the necessary wire runs. Cables with similar noise susceptibility levels and noise generation power levels can then be grouped in trays and conduit, and intermixing is thereby avoided.

6.4.3.1 Wiring Levels and Classes. There are four basic levels or classes of wiring which can be identified. These classes or levels are listed below with typical signals defined for each level.

6.4.3.1.1 Level 1 — High Susceptibility. Analog signals of less than 50 V and digital signals of less than 15 V.

(1) Common returns to high-susceptibility equipment
(2) Control common tie (CCT)
(3) Dc power supply buses feeding sensitive analog hardware
(4) All wiring connected to components associated with sensitive analog hardware (for example, strain gauges, thermocouples, etc)
(5) Operational amplifier signals
(6) Power amplifier signals
(7) Output of isolation amplifiers feeding sensitive analog hardware
(8) Phone circuits
(9) Logic buses feeding sensitive digital hardware
(10) All signal wires associated with digital hardware

6.4.3.1.2 Level 2 — Medium Susceptibility. Analog signals greater than 50 V and switching circuits.

(1) Common returns to medium-susceptibility equipment
(2) Dc bus feeding digital relays, lights and input buffers
(3) All wiring connected to input signal conditioning buffers
(4) Lights and relays operated by less than 50 V
(5) Analog tachometer signals

6.4.3.1.3 Level 3 — Low Susceptibility. Switching signals greater than 50 V, analog signals greater than 50 V, regulating signals of 50 V with currents less than 20 A, and ac feeders less than 20 A.

(1) Fused control bus 50—250 V dc
(2) Indicating lights greater than 50 V
(3) 50--250 V dc relay and contactor coils
(4) Circuit breaker coils of less than 20 A
(5) Machine fields of less than 20 A
(6) Static master reference power source
(7) Machine armature voltage feedback
(8) Machine ground-detection circuits
(9) Line-shunt signals for induction
(10) All ac feeders of less than 20A
(11) Convenience outlets, rear panel lighting
(12) Recording meter chart drives
(13) Thyristor field exciter ac power input and dc output of less than 20 A

6.4.3.1.4 Level 4 — Power. ac and dc buses of 0—1000 V with currents of 20—800 A.

(1) Motor armature circuits
(2) Generator armature circuits
(3) Thyristor ac power input and dc outputs
(4) Primaries and secondaries of transformers above 5 kVA
(5) Thyristor field exciter ac power input and dc output with currents greater than 20 A
(6) Static exciter (regulated and unregulated) ac power input and dc output
(7) 250 V shop bus
(8) Machine fields over 20 A

6.4.3.2 Class Codes. Within a level, conditions may exist that require specific cables, and regrouping is not allowed. This condition may be identified by a class coding system similar to the following:

A analog inputs, outputs
B pulse inputs
C contact and interrupt inputs
D decimal switch inputs
E output data lines
F display outputs, contact outputs
G logic input buffers

S special handling of special levels may require special spacing of conduits and trays, such as signals from commutating field and line resistors, or signals from line shunts to regulators, or power > 1000 V or > 800 A, or both

U high-voltage potential unfused greater than 600 V dc

6.4.3.3. Tray Spacing. Table 7 indicates the minimum distance in inches between the top of one tray and the bottom of the tray above, or between the sides of adjacent trays. This also applies to the distance between trays and to power equipment of less than 100 kVA.

6.4.3.4 Tray–Conduit Spacing. Table 8 indicates the minimum distance in inches between trays and conduits. This also applies to the distance between trays or conduits and power equipment of less than 100 kVA.

6.4.3.5 Conduit Spacing. Table 9 indicates the minimum distance in inches between the outside surfaces of conduits being run in banks. This also applies to the distance between conduits and power equipment of less than 100 kVA.

6.4.3.6 Notes on Tray and Conduit Cabling.

(1) Levels 3 and 4 may be run in a common tray, but should be separated by a barrier. Spacing should be level 4. (This barrier does not necessarily have to be grounded.)

(2) When separate trays are impractical, levels 1 and 2 may be combined in a common tray, provided the levels are separated by a grounded steel barrier. This practice is not as effective as separation, and some rerouting at installation may be required. When levels 1 and 2 are run side by side in trays, a 1 in minimum spacing is recommended.

(3) When unlike signal levels must cross either in trays or in conduits, they should cross at 90° angles at a maximum spacing. Where it is not possible to maintain spacing, a grounded steel barrier should be placed between unlike levels at the crossover point.

(4) Trays containing level 1 and 2 wiring should have solid bottoms. Ventilation slots or louvers may be used on other trays. Tray covers on level 1 and 2 trays must be used to provide complete shielding. Cover contact to side rails must be positive and continuous to avoid high-reluctance air gaps which impair shielding. Trays for all levels should be metal, solidly grounded with good ground continuity.

(5) Trays and conduits containing levels 1, 2, and 3S should not be routed parallel to high-power equipment enclosures of 100 kVA and larger at a spacing of less than 5 ft for trays and 2½ ft for conduit.

(6) Where practical for levels 4 and 4S wiring, the complete power circuit between equipment should be routed in the same path, that is, the same tray or conduit. This practice will minimize the possibility of power and control encircling each other.

(7) When entering terminal equipment and the spacings listed above are difficult to maintain, parallel runs should not exceed 5 ft in the overall run.

(8) All spacing given in Table 8 assumes that the level 1 and 2 trays will

Table 7
Tray Spacing (inches)

Level	1	2	3	3S*	4	4S*
1	0	See 6.4.3.6 (2)	6	6	26	26
2	See 6.4.3.6 (2)	0	6	6	18	26
3	6	6	0	0	See 6.4.3.6 (1)	12
3S*	6	6	0	0	8	18
4	26	18	See 6.4.3.7 (1)	8	0	0
4S*	26	26	12	18	0	0

*See 6.4.3.2, class code S.

Table 8
Tray–Conduit Spacing (inches)

Level	1	2	3	3S*	4	4S*
1	0	1	4	4	18	18
2	1	0	4	4	12	18
3	4	4	0	0	0	8
3S*	4	4	0	0	6	12
4	18	12	0	6	0	0
4S*	18	18	8	12	0	0

*See 6.4.3.2, class code S.

Table 9
Conduit Spacing (inches)

Level	1	2	3	3S*	4	4S*
1	0	1	3	3	12	12
2	1	0	3	3	9	12
3	3	3	0	0	0	6
3S*	3	3	0	0	6	9
4	12	9	0	6	0	0
4S*	12	12	6	9	0	0

*See 6.4.3.2, class code S.

not be left uncovered; otherwise Table 7 spacing must be used.

(9) Where 0 is indicated as a tray or conduit spacing, the levels may be based on the worst condition.

(10) Spacing internal to equipment and at its terminal boards may be different from external spacing required in trays and conduits.

6.4.3.7 Notes on Specific Routing of Levels

6.4.3.7.1 Pullboxes and Junction Boxes

(1) Within the confines of pull boxes and junction boxes, levels should be kept separate, and grounded barriers should be used to assist level spacing.

(2) Tray-to-conduit transition spacings and separations are potential sources of noise. Care should be taken to cross unlike levels at right angles and maintain required separations. Transition areas should be protected in accordance with the level recommendations.

6.4.3.7.2 Control Pulpit Cabling.

(1) Where random floor wiring is absolutely necessary, unlike levels should be separated and like levels tied together or contained by some suitable means, and basic level rules followed.

(2) A tray network in the floor provides better noise-free installation than a random wired floor. These trays will provide better shielding and assure proper separation if properly planned and installed.

(3) The same degree of separation is required in the routing of cables in ceiling and walls from overhead cabinets and wall-mounted control units.

6.4.3.7.3 System Additions and Modifications

(1) The reduction of noise requires careful planning so that proper separation is maintained and that lower and higher levels never encircle each other or run parallel for long distances

(2) The use of existing conduits or trays is practical as long as the level spacing can be maintained for the full length of the run.

(3) Where possible, level 1 and 2 cables should be routed in a path apart from existing cables as these existing cables are generally of a high-voltage potential and noise-producing variety.

(4) Barriers should be used in existing pull-boxes and junction boxes for low-level wiring so that the possibility of noise pickup is minimized.

(5) Care should be taken so that low-level signals are not looped around high control or power level conduits or trays.

6.4.3.7.4 Conduits around and through Machinery Housings

(1) Care should be taken to plan level spacings on both embedded and exposed conduit in and around machinery. Runs containing mixed levels should be minimized to 5 ft or less in the overall run.

(2) Conduits running through and attached to machinery housings should follow level spacing recommendations and should be discussed with the machinery builder early in the project.

(3) Trunnions entering operator cabinets should be kept as short as possible, thereby minimizing parallel runs. Unlike levels should be kept separate for as long a distance as possible prior to entering the equipment.

(4) Where different levels are running together for short distances, each level should be contained by cord ties, barriers, or some logical method so that intermixing is avoided.

6.4.3.7.5 Transitional Areas

(1) When entering or leaving conduits or trays, care should be taken to ensure that cables of unlike levels do not become intermixed.

(2) The use of grounded steel barriers for level separation may be advisable when parallel runs over 5 ft overall become necessary.

6.4.3.7.6 Level Identification at Installation. The application of levels, if properly implemented, can be monitored and identified for future maintenance and for controlling level practices at installation. One method of implementing level control is to color code the outer jackets of the interconnecting cables. For example: red and black could be used for levels 4 and 4S, grey and white for levels 1 and 2, and possibly blue for level 3. The main purpose is to make an easily recognizable means of keeping cables segregated. Exceptional care must be taken, however, to ensure that cable coloring is not violated. In addition, all conduits and trays should have level identification by coding, numbers, or some other means which could be used to identify conduits or trays at junction points or at periodic intervals. Cable level identification may be implemented as color-coded tags at the termination ends of the cables. Either the above or some other comparable system will assist the user in the application of levels.

6.4.4 Impedance Coupling (see also 3.4.1). Impedance coupling, as used in this guide, refers to a form of electrical noise coupling whereby currents flow through an impedance *common* to several circuits and thus produce a voltage drop across the common impedance such that the impedance-coupled voltage drop adds to or subtracts from the desired signal voltages. The objective of this section is to describe cable-interconnecting systems for controllers which will minimize or eliminate impedance-coupled noise problems. The two basic methods of combating impedance-coupled noise problems are as follows:

(1) Make the wire size of a signal wire which is shared by more than one signal path (common signal return) as large as practical to reduce the impedance of the wire. This reduces the voltage drop produced by the noise currents and thus reduces the possibility of malfunction of the controller.

(2) Replace the wire shared by more than one signal path (common signal return wire) with individual signal return wires.

The following subsections describe various cable-interconnecting systems which can be utilized between the controller and external transmitters and receivers to eliminate or minimize impedance-coupled noise. The external transmitters and receivers may be located in another controller, as depicted by the dashed boxes around the transmitters T and receivers R on the left-hand sides of Figs 84-86, or they may be individual external control elements located at scattered locations, such as on motors or on housings of machine tools. Typical transmitters T could include control elements, such

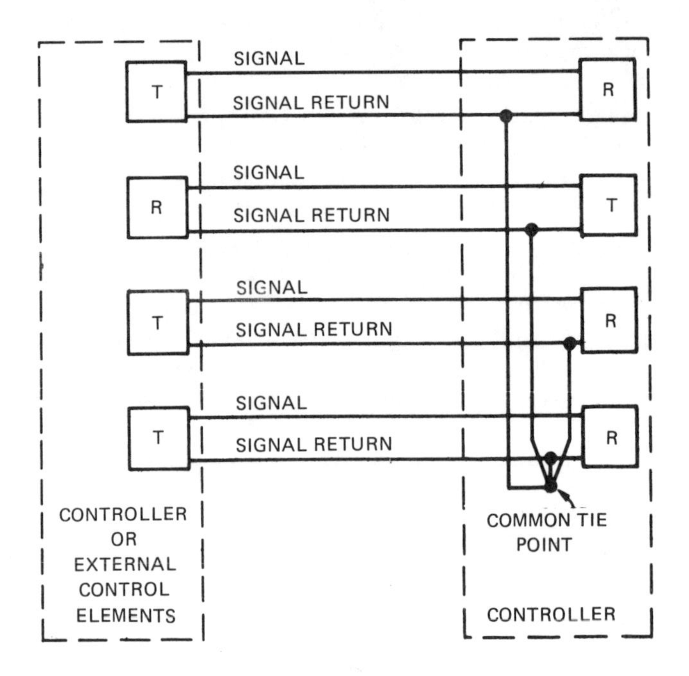

Fig 84
Cabling System Illustrating Individual Signal Returns

as limit switches, thermocouples, tachometer pickups, or potentiometers. Typical receivers R could include operational amplifiers, solenoids, relays, recorders, and digital logic circuits.

6.4.4.1 Industrial Signal Return. Where only a few signal wires are involved, it is recommended that individual wires for return signals be used. Such a system will virtually eliminate impedance-coupled noise at the expense of nearly doubling the number of wires in the interconnecting system. In Fig 84 each transmitter is first separately wired to its respective receiver, and then a wire is run from the signal return to a common tie point. In this system the common conductor is reduced to a terminal point (common tie point) rather than a wire, and thus no signal return current flows through any other signal return wire. It should be noted that a common tie point may be inherent in the system if there exists a common dc power supply for two or more of the transmitters T and receivers R, or two or more signals share a common source. This type of wiring cannot be relied upon for communication between any transmitter or receiver and any other transmitter or receiver since this would result in the signal current flowing through two or more signal returns.

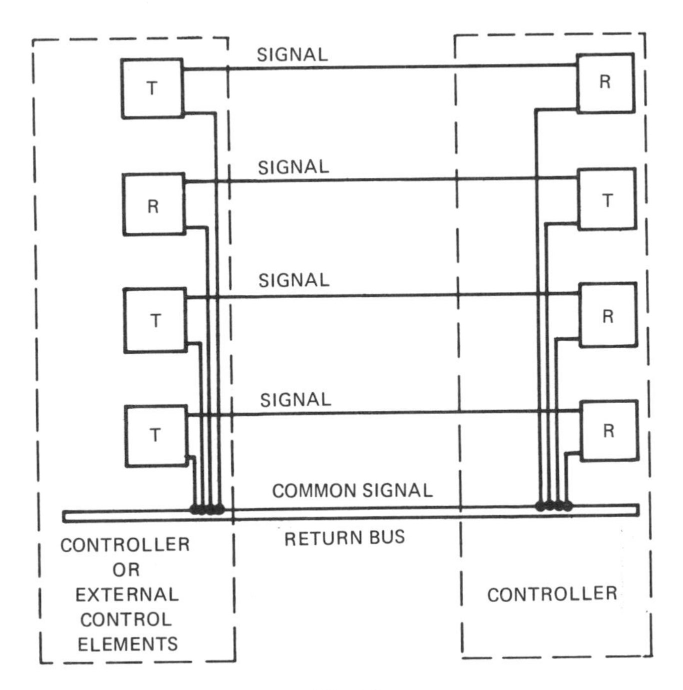

Fig 85
Cabling System Illustrating Common Signal Return

6.4.4.2 Common Signal Return. Where many signal wires are involved it is recommended that the common return signal wire be made a cable or a bus. This common return should be made as heavy as practical so that the designer will have almost complete freedom in routing signal wires. This method, as shown in Fig 85, minimizes the number of interconnecting cables and should be used when a system of individual signal return cables would involve too complicated a network of signal return wires to be implemented reliably and at reasonable cost. This is particularly true when it is necessary for any transmitter or receiver to communicate with any other transmitter or receiver.

6.4.4.3 Combined Individual and Common Signal Returns. For some systems it may only be practical to make the common signal return cable large enough to carry only a portion of the signal return currents. In this case a combination of individual signal returns and a common signal return, as shown in Fig 86, can be employed. The system should be designed such that the most sensitive signal wires to equipment with relatively high noise susceptibility are identified to employ an individual signal return wire.

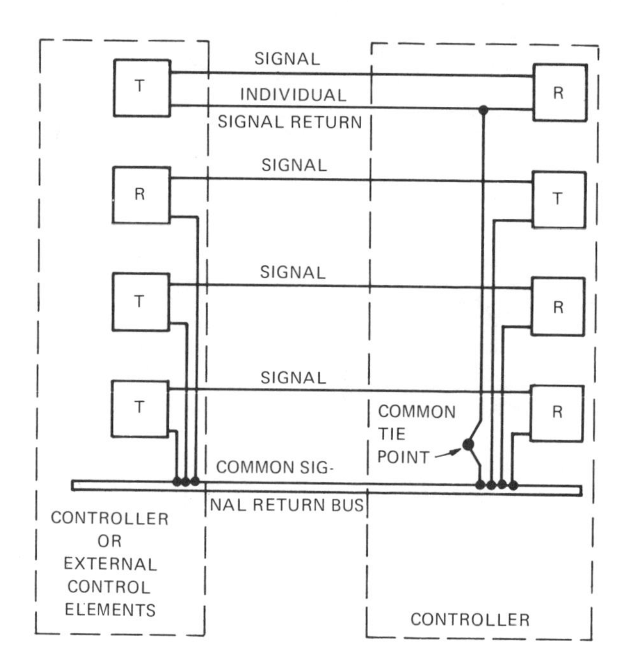

Fig 86
Cabling System Illustrating Combined Individual and Common Signal Returns

6.5 Suppressive Techniques. This section deals with the suppression of electrical noise at its source and filtering.

6.5.1 Suppression of the Source. When a source of electrical noise is known or suspected, appropriate action may be taken to reduce or eliminate the noise. Some methods for dealing with the most troublesome noise sources are listed in this section.

This method of dealing with electrical noise may be applied after installation and during checkout of the system if it is found to be necessary for proper control function. If suppressive techniques are used, precautions should be taken to prevent further control expansion or changes from incorporating similar noise sources which have not been suppressed.

The voltage breakdown of conductor spacing through air or over insulation surface will be the maximum voltage that can exist at specific locations in a system. While it is not a recommended method, voltage breakdown between live conductors or between a live conductor and ground will partially suppress transient voltages which would otherwise be greater than this breakdown value.

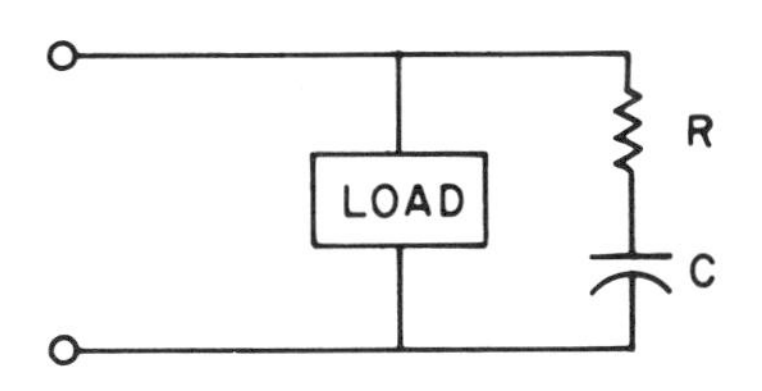

Fig 87
Circuit Showing Resistor in Series with Capacitor

6.5.1.1 Suppression of the Showering Arc. Suppression of electrical noise due to the showering arc (see 4.1.1) can be reduced or eliminated by several techniques. The best technique depends on the particular application and the amount of suppression desired.

6.5.1.1.1 Capacitive Quenching. A capacitor of proper size connected in parallel with the inductive load can completely eliminate the showering arc. A resistor (preferably noninductive) is usually connected in series with the capacitor to eliminate large inrush currents on circuit closure.

For ac or dc relays, contactors, and solenoids, use a 220 Ω resistor in series with a 0.5 μF 1000 V capacitor and connect as shown in Fig 87. It may be more economical from the standpoints of cost or size to use an appropriate voltage limiter (see 6.5.1.1.2) in parallel with the capacitor and resistor. The capacitor voltage rating could then be based upon the voltage limiter voltage rating.

The power which will be dissipated in the resistor when ac circuits are energized is shown in Table 10.

6.5.1.1.2 Voltage Limiters. The voltage amplitude of showering arcs will ordinarily reach 2000 V. Any kind of device or circuit that will limit the voltage to some smaller value will reduce the electrical noise.

If the voltage can be limited to 300 V or less, the noise due to showering arcs will be eliminated. See Fig 88 for types of limiters. A resistor in parallel with the load will limit the voltage to V_{max}, where

$$V_{max} = IR$$

I being the maximum load current.

In dc circuits the power dissipation under steady-state conditions can be

Table 10
Power Dissipated in Resistor

Power Supply	Resistor Power (W)
115 V	
25 Hz	0.019
50 Hz	0.076
60 Hz	0.11
230 V	
25 Hz	0.076
50 Hz	0.31
60 Hz	0.44
460 V	
25 Hz	0.31
50 Hz	1.2
60 Hz	1.8
550 V	
25 Hz	0.42
50 Hz	1.7
60 Hz	2.4

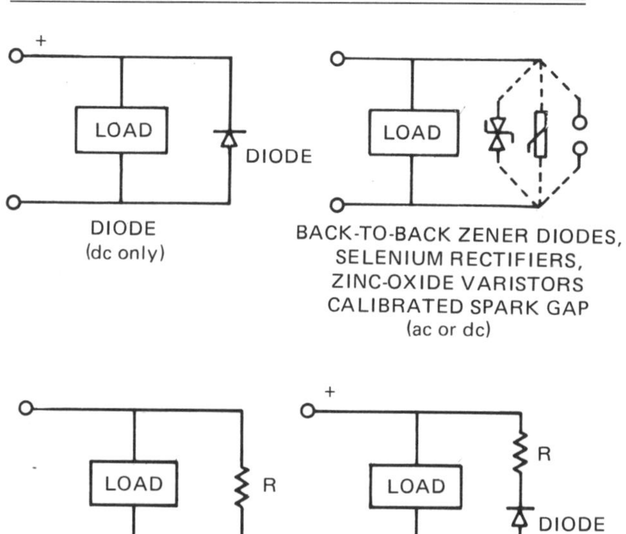

Fig 88
Types of Limiters

eliminated by placing a diode in series with the resistor. Maximum suppression in dc circuits can be obtained by using a diode only.

Back-to-back zener diodes, selenium rectifiers, or zinc-oxide varistors are very effective for ac or dc circuits.

A combination of the circuits described may be used. A voltage limiter such as back-to-back zener diodes or zinc-oxide varistors may be used to limit the voltage peak occurring with a capacitive quench circuit.

CAUTION: The use of these noise suppression networks can modify the speed of operation of the load. The use of a diode only on a dc contactor or relay may increase the dropout time by a factor of 10 or 20. For this reason diodes and resistor diode networks must be used with caution on brakes, clutches, motor fields, etc, where the speed of operation may affect the safety of operation.

6.5.1.2 SCR Phase Control and Rectifiers. The electrical noise caused by the large rate of change of current and voltage due to phase-control devices (SCRs, triacs, etc) can be reduced by using appropriate inductors in series with the switching device. This practice may then also require a series resistor–capacitor network in parallel with the switching device to supply the required device holding current until the load current builds up to this value. (Resistive load controllers should use zero point switching rather than phase control wherever possible; heater loads, for example.)

6.5.1.3 Transient Overvoltages. Lightning arresters selected according to power-line rated voltage and connected from each power line to ground, preferably at the building service entrance, should be used where transient overvoltages due to lightning are a problem. The arrester short-circuits the line to ground for at least a half-cycle in performing its protective function.

Voltage limiters (see 6.5.1.1.2) may be connected across the power line but must be selected with sufficient power rating to withstand the transient voltage damage. This type of protection permits continued operation of the devices connected to the power line.

Protection against catastrophic failure can be provided by interrupting power to the device when a voltage sensor detects an overvoltage transient. The required large speed of response can be obtained by detecting the transient with a zener diode which "turns on" an SCR to short-circuit the power line. Power-line circuit breakers, fuses, or current-limiting devices may then function to prevent damage due to the short circuit. This type of protection is called *crowbar*.

6.5.2 Filtering

6.5.2.1 General Installation Requirements.

(1) The filter should be placed as close to the source of noise as possible to prevent it from contaminating several receptors.

(2) If the noise contains high-frequency components, use feed-through capacitors or filters (fed through shield):

(a) In the case of capacitors, to prevent detrimental effects of capacitor leads in the shunt branch. (The capacitor becomes inductive beyond the resulting series resonance.)

(b) To prevent input–output coupling, that is, bypassing of the capacitor or filter. If no shield is available shield the input line to filter and, if necessary, also the output line.

(c) To prevent destruction of the filter select the proper rating (voltage, current, duty cycle, peak interference, etc.)

6.5.2.2 Selection of the Most Appropriate Filter. The character and source of the noise can be determined with the help of an oscilloscope (for voltage measurements) plus a small series resistor or current transformer (for current measurements) and by judiciously disconnecting or shielding lines.

Figure 89 illustrates schematically the most often encountered noises. In the discussion to follow, each generic noise will be briefly described, and the most appropriate filtering will be listed, with the most economic approach given first. (P will stand for power feed line, dc or ac, C for control line, and S for signal line.)

(1) Figure 89(a1)-(a3) [P, 60 Hz). The ac line voltage fluctuates [Fig 89(a1)], or the half-cycle is missing, for example, due to lightning arresters [Fig 89(a2)], or highly energetic notching occurs [Fig 89(a3)]. For all these conditions a simple rugged remedy exists. The most economic means to get rid of these difficulties is an improved ferroresonant transformer. (A conventional ferroresonant transformer is insufficient in filtering and collapses under strong notching and half-cycle interruptions.) If the harmonic content (5-10%) of such improved ferroresonant filter regulators is not acceptable, more costly solutions such as inverters, are to be used [11].

(2) Figure 89(b) [P, dc]. For the elimination of fluctuations or notching in dc supply lines, a storage capacitor is the first device to be tried. For excessive, particularly random, current pulses, active power-line filters, somehow modified regulators, are necessary [12].

(3) Figure 89(c) [P, C, S]. Spikes exceeding the extreme values of the voltage under consideration are most inexpensively reduced by nonlinear filters, also called limiters or slippers. [If spikes occur within the ac boundaries, go back to (1).] For power feed lines, spark gaps and sharply bent selenium rectifiers, for more limited power, varistors (zinc oxide), and for low power, as in operational amplifiers, back-to-back diodes are quite effective [14].

(4) Figure 89(d) [P, C, S]. If on the oscilloscope the interference is clearly indicated as having higher frequencies than the power control or signal frequency, low-pass filtering is the answer. The simplest approach is the insertion of filter capacitors (feed-through) across the line. Particular caution is necessary in the case of tantalytic capacitors for dc lines. These capacitors lose much of their effectiveness at higher frequencies [7]. For very high frequencies, ferrite-loaded ceramic filter capacitors [7], or ferrite beads for low-impedance loads, are most effective.

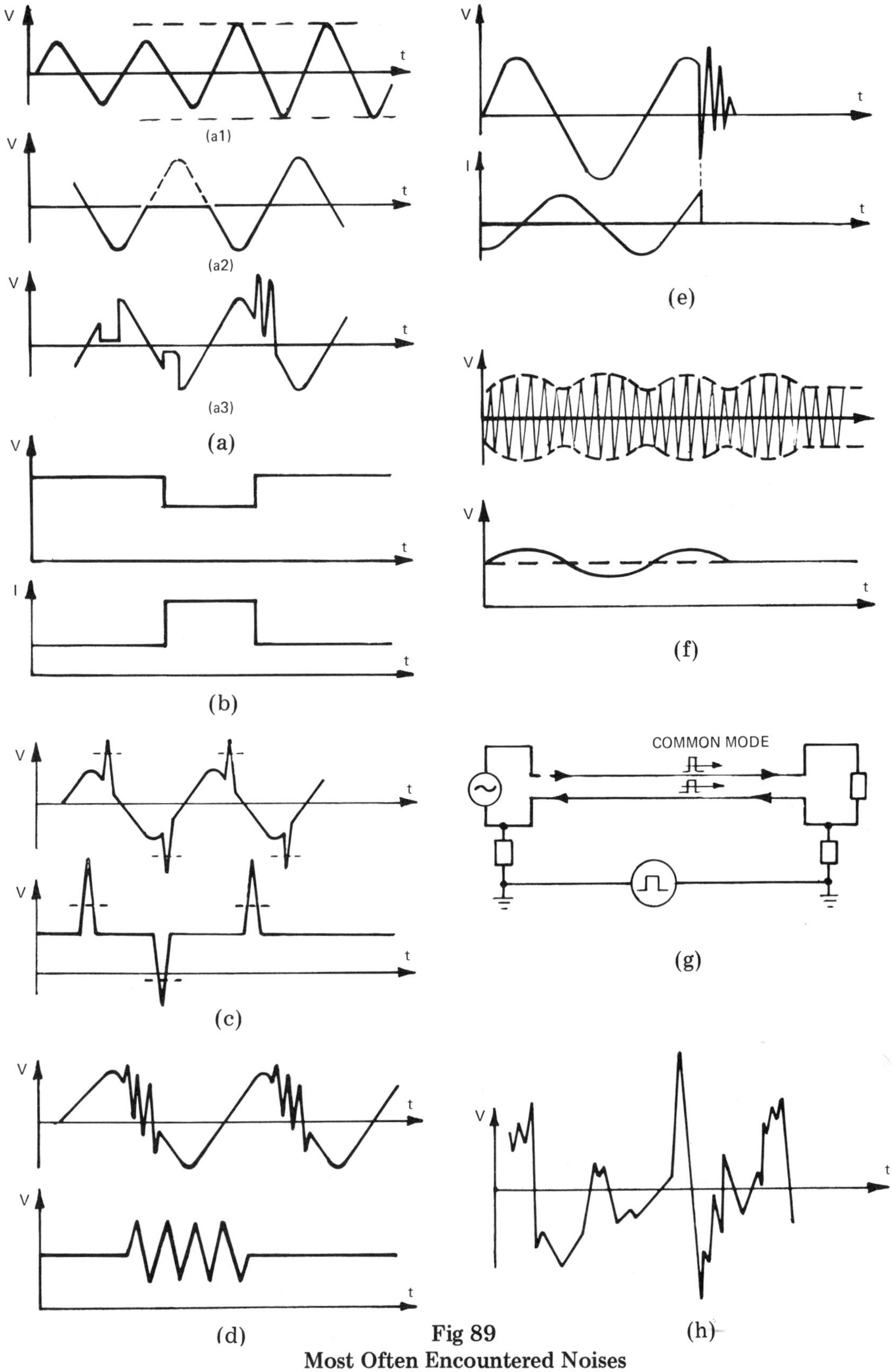

Fig 89
Most Often Encountered Noises

If higher insertion loss is required (RC filters for signal lines), properly rated LC interference filters should be applied. For ac lines the π filter should be preferred; for dc lines the T filter is mostly more effective. Also, if one knows something about the load impedance, one will prefer T filters for low impedances (at the critical noise frequencies) and π filters for high impedances. L filters may often be the cheapest, but they have the strongest tendency to ring [10], [13] [See (5).]

(5) Figure 89(e) [P, C]. The ringing of filters, quite often pronounced in switching transients, or the existence of a high harmonic content in the feed-through power, control, or signal line, resulting in insertion gain of the filter (mostly in the so-called passband), can cause damage to the filter or can, for instance, cause misfiring of triacs or SCRs. In such cases it is best to select T filters which have the least tendency to ring (except for capacitance loads for ac lines). If this does not help, lossy filters are to be applied (see 4.5.2.2 or [8].

(6) Figure 89(f) [S]. If a strong 60 Hz signal is superimposed on ac or dc signals, twin-T RC networks quite often suffice to get rid of the 60 Hz interference. In critical cases active low-pass filters may be necessary.

(7) Figure 89(g) [P, C, S]. Common mode (in contrast to normal mode where a forward and a return wire are involved) is encountered when a wire pair acts essentially as one conductor, and the return path is through some other conductor, mostly the ground return. In such cases conventional normal-mode filters do not do much good. However, before introducing common-mode filters (small, with bifilarly wound inductors; no biasing effect), in particular when signal and control lines are involved, the ground return should be suppressed by using simpler means. More and more optoelectronic isolators (with CdS sensors to slow normal-mode transients or with silicon sensors for fast normal-mode operation) are employed to prevent ground loops and, thus, the formation of common-mode operation. Baluns (bifilar coils) for low-frequency common-mode interference, and ferrite cores for high-frequency interference and low-impedance common-mode situations, also are often helpful. They do not provide isolation. Only in cases of extreme interference will one apply common-mode filters or isolation transformers.

(8) Figure 89(h) [S]. If signals are obscured by excessive noise, and if no amount of conventional frequency-separating filtering works, since the power spectra of noise and signal overlap, averaging, correlation, matching (to the signal), or phase-locked loop techniques will permit extraction of the signal out of the noise [7].

6.6 Reduction of Interference [4], [9]. The techniques for the reduction of electromagnetic interference are sometimes critically dependent upon the frequency range of interfering sources. The interfering sources are, in broad terms, classified as low- and high-frequency sources without a specified line of demarcation. Depending upon the specific application, frequencies below the range of 100 kHz to 1 MHz are classified as low frequencies. For applications to cable shielding and grounding, the frequencies which have their quarter-wavelengths greater than the cable lengths are classified as low frequencies.

High-frequency interference sources in electrical control systems may be divided into three classes: (1) continuous, for example, pulse-width-modulated (PWM) sources, or thyristor-controlled choppers, (2) pulse trains, for example, showering arcs from switch and relay operations, and (3) single high-power pulses, for example electromagnetic pulses (EMP) and lightning. Because the high-frequency interference source cannot be suppressed in most cases, means must be developed either to attenuate the high-frequency interference during transmission or to decrease the susceptibility of the receiver.

The transmission may take place via either conduction or radiation, or both. Sometimes the transmission may take place over *sneak* paths which may be very difficult to identify. The signal transmission line must be shielded to reduce radiative pickup of high-frequency interference. For low-frequency electrical noise the shield must be grounded at one point to avoid ground-loop currents. For high-frequency interference one-point ground is not only meaningless but even harmful. High-frequency currents do flow to ground at the *isolated* end of the shield through capacitive coupling. Moreover, the shield may act as an antenna, radiating the high-frequency interference. Therefore the shield must be grounded at multiple points. For most control systems twisted, shielded pairs of cables are most suitable where the shield is connected to ground at multiple points. In high-frequency applications coaxial cables are recommended. Where the interfering source is strong or the protected equipment is sensitive to interference, two shields (triaxial cable) must be used, the inner shield having one-point grounding and the outer shield having multiple grounding points (Fig 80). The impedance of a shield should be the minimum possible. Metal hose with helical turns of metal wrap spaced with a cotton or fiber cord constitutes a solenoid and gives large inductive reactance at radio frequencies with essentially no shielding action. Knitted wire mesh tubing also produces poor results due to its high ohmic resistance as well as porosity to high-frequency penetration. Solid foils of copper and aluminum provide the best shielding at high frequencies. Proper routing also helps to attenuate high-frequency interference. Where possible the cables should be laid as near to the ground surface as possible to reduce the radiating loop.

Common-mode-rejecting inductors are extremely useful in reducing ground-loop currents (Fig 59). These inductors are formed by winding a few turns of the signal cable through a toroidal magnetic core. These inductors are so useful that they have been incorporated in many systems as normal operating procedure.

Isolation transformers (Fig 82) are useful in suppressing high-frequency interference during transmission. Isolation transformers have also been used with advantage in ac power supplies to suppress high-frequency interference coming from the supply lines. However, isolation transformers have often caused more problems when not selected with caution. Such transformers must have very low inter-winding capacitance (on the order of picofarads) and no conductive path between the grounded terminals of either winding. Many commercially available isolation transformers have the grounded terminals of both windings connected solidly. Care should be taken to disconnect such connections. It may, however, violate the prescribed safety rule.

Optoelectronic couplers are also used to suppress high-frequency interference (Fig 83). However, it should be borne in mind that any high-frequency interference induced in the transmission line from the signal source to the emitter (for example, electrical-to-optical converter) will be converted to light output and passed on to the receiver. The most appropriate solution will be to replace the entire electrical transmission line by fiberoptic transmission. This mode of signal transmission is being increasingly used because of its lower losses (less than 20 dB/km) and competitive cost.

To reduce the susceptibility of a receiver, the first line of defense will be to select slow-response components where possible. For instance, RC networks in control and signal lines are highly effective. The lower the cutoff frequency of such a filter, the slower will be the response. To reduce the destructive nature of switching and lightning surges, nonlinear resistors are the best surge arresters. Metal-oxide varistors, avalanche diodes, and silicon-carbide (thyrite) and selenium-oxide devices have been effectively used as nonlinear resistors. The nonlinear resistor should be selected such that (1) its leakage current is negligible within the operating limits of the signal voltage levels, (2) the voltage drop across the nonlinear resistor under the surges is lower than the safe transient withstand voltage level of the receiver or its individual components, and (3) the nonlinear resistor is not destroyed by the energy absorbed from the surge. The first condition is always easy to meet, because the operating limits of the signal voltage levels are or should always be known. The second condition is difficult to meet if the safe transient withstand voltage level of the receiver is not known. In that case the equipment should be derated, and its maximum steady-state voltage limit should be used instead. The third condition is the most difficult to achieve, because the energy of the random surge is seldom known. In such cases the nonlinear resistor with the maximum available energy-handling capability which is permitted by cost and space considerations must be used. This will improve reliability, although nondestruction is not 100% ensured.

The surge arrester must be installed as close to the protected apparatus as possible to derive the best benefit of surge protection. The lead length of the surge arrester must also be short to minimize the inductive voltage drop.

The susceptibility of a receiver is also improved by proper shielding. A ferromagnetic material generally provides better shielding than copper or aluminum, even if its effective permeability is low at high frequencies. However, copper or aluminum is normally chosen for practical reasons (such as weight). Multiple shielding should be used for best results in extreme conditions, with the copper layer toward the source (outside) and a ferromagnetic material as the inside layer.

Permanent joints in the shield should preferably be welded. If riveted, the rivets should be very close and so tight that corrosion and oxidation cannot penetrate into the originally clean metallic interface. For doors, double rows of finger stock of noncorrosive spring material should be used. Holes in the shield should be as small in dimension as possible, if they cannot be avoided altogether. The rounding of edges and corners is also recommended.

In the application of shielding, whether for an apparatus or for a transmission line, the best results are achieved when the shield encloses the apparatus (or transmission line) entirely, and when the shield material is solid (not meshes). The ground lead of the shield must be short and flat to introduce as low an inductance as possible to the ground current.

6.7 References

[1] ANSI/NFPA 70-1981, National Electrical Code.

[2] IEEE Std 142-1982, Recommended Practice for Grounding of Industrial and Commercial Power Systems (IEEE Green Book).

[3] BEADLE, R.G., JARVININ, W.B., BLOODWORTH, T.H., GEISHEIMER, F., and BRISLAND, D.W. Reduction of Electrical Noise in Steel Mills. *IEEE Transactions on Industry and General Applications*, vol IGA-3, Mar/Apr 1967.

[4] BIRKEMEIER, W.P. Mitigation of Radio Frequency from PWM Servo Drives. *Proceedings of the 1979 IEEE Industry Applications Society Annual Meeting*, pp 597–599.

[5] GEISHEIMER, F. Recommendations of an Engineering Contractor. *IEEE Transactions on Industry and General Applications*, vol IGA-3, Mar/Apr 1967, pp 83–87.

[6] KLIPEC, B.E. Reducing Electrical Noise in Instrument Circuits. *IEEE Transactions on Industry and General Applications*, vol IGA-3, Mar/Apr 1967.

[7] SCHLICKE, H.M. *Practical Design Electromagnetic Compatibility*. FICCHI, R.F. (Ed). New York: Hayden Book Co, 1971.

[8] SCHLICKE, H.M. Assuredly Effective Filters. *IEEE Transactions on Electromagnetic Compatibility*. Vol EMC-18, no 3, Aug 1976, pp 106-1100.

[9] SCHLICKE, H.M. *Electromagnetic Compossibility*. (Applied Principles of Cost Effective Control of Electromagnetic Interference and Hazards), 2nd Enlarged Ed. New York: Marcel Dekker, 1982.

[10] SCHLICKE, H.M., BINGENHEIMER, A.J., and DUDLEY, H.S. Elimination of Conducted Interference, A Survey of Economic, Practical Methods. Invited paper at the 1971 Annual Meeting of the IEEE Industry and General Applications Group.

[11] SCHLICKE, H.M., FREDRICKSON, R., and VEBBER, W. Filter-Regulators for 60 Hz Power Lines. *Record of the IEEE International EMC Symposium*, Philadelphia, PA, 1971.

[12] SCHLICKE, H.M., and WEIDMAN, H. Compatible EMI Filters. *IEEE Spectrum*, vol 4, Oct 1967, pp 59-68.

[13] SCHLICKE, H.M., and WEIDMAN, H. Effectiveness of Interference Filters in Machine Tool Control. Presented at the 19th Annual IEEE Machine Tool Conference, Detroit, MI, Oct 1969.

[14] WILLARD, F.G. Transient Noise Suppression in Control Systems. *Control Engineering*, vol 17, Sept 1970, p 59.

Index

INDEX

INDEX

T

U

V

W

Z